# 金圆规奖
# 引领意大利设计潮流
# 七十年

意大利工业设计协会博物馆
上海久事美术馆　编

# Compasso d'Oro
# Award

# Seventy Years Leading
# Italian Design Trends

上海书画出版社　ꝨTRECCANI

© ISTITUTO DELLA ENCICLOPEDIA ITALIANA
FONDATA DA GIOVANNI TRECCANI S.P.A., ROMA 2024

*President*
Franco Gallo

*Vice Presidents*
Domenico Arcuri, Giovanni Puglisi

*Director General*
Massimo Bray

*Head of international projects*
Giovanna Fazzuoli

*Editorial coordination*
Cristina Bravi

*Editing*
Emma Allegretti


ISTITUTO ITALIANO DI CULTURA DI SHANGHAI

*Director*
Francesco D'Arelli

Ruirui Chu, Adele Lobasso, Marco Lovisetto, Qiling Yue


ADI DESIGN MUSEUM

*President ADI*
Luciano Galimberti

*President Fondazione ADI*
Umberto Cabini

*Director*
Andrea Cancellato

*General Coordinator*
Maria Pina Poledda

*Edited by*
ADI Design Museum
Shanghai Jiushi Art Museum

*Curated by*
Maite García Sanchis *with*
Francesca Balena Arista
Giovanni Comoglio

*Exhibition design by*
Cibic Workshop, Aldo Cibic and Joseph Dejardin

*Exhibition photo by*
©RAWVISION studio

# 引领潮流七十年
## ——意大利金圆规奖经典设计作品展

外滩 18 号久事艺术空间
2023.11.9—2024.3.31

# Compasso d'Oro Award:
## Seventy Years Leading Italian Design Trends

Bund 18, Shanghai
9 November 2023 - 31 March 2024

---

**支持与推动者**
意大利驻沪总领事馆
安缇雅 意大利驻沪总领事

意大利驻沪总领事馆文化处
达仁利 意大利驻沪总领事馆文化处处长

意大利特雷卡尼百科全书研究院
马西莫 · 布莱 总经理

ADI 设计博物馆
卢恰诺 · 加林贝蒂 ADI 主席
翁贝托 · 卡比尼 ADI 基金会主席
安德烈 · 坎切拉托 ADI 设计博物馆馆长
玛丽亚 · 皮娜 · 波莱达 ADI 设计博物馆总协调员

上海久事美术馆
谷际庆 久事美术馆馆长
王雯洁 久事文传董事长

**展览团队**

**项目总监**
胡喆

**展览策展**
梅特 · 加西亚 · 桑奇斯
乔万尼 · 科莫利奥
弗朗西斯卡 · 巴勒娜 · 阿里斯塔
凌敏

**展览研究**
亚历山德拉 · 丰塔内托
斯特凡妮 · 迪 · 玛丽亚

**展览统筹**
陈轩凯
陈熙亮
成睿文
索菲亚 · 罗德里格斯

**展览设计**
西比克工作室

**视觉设计**
彭浩

**媒体**
庞蔚然
曹洋琪

**公共教育**
邹卓尔
关玥

**Supported and Promoted by**
**Consolato Generale d'Italia a Shanghai**
**Tiziana D'Angelo, Consul General**

Istituto Italiano di Cultura di Shanghai
**Francesco D'Arelli, Director**

Istituto dell'Enciclopedia Italiana-Treccani
**Massimo Bray, Director General**

ADI Design Museum
**Luciano Galimberti, ADI President**
**Umberto Cabini, ADI Foundation President**
**Andrea Cancellato, ADI Design Museum Director**
**Maria Pina Poledda, General Coordinator ADI Design Museum**

Shanghai Jiushi Art Museum
**Gu Jiqing, Shanghai Jiushi Art Museum Director**
**Wang Wenjie, Chairman of the Board of JUCE**

Exhibition Team

Project Director
**Hu Zhe**

Exhibition Curator
**Maite García Sanchis**
**Giovanni Comoglio**
**Francesca Balena Arista**
**Ling Min**

Research
**Alessandra Fontaneto**
**Stefania Di Maria**

Exhibition Coordinators
**Chen Xuankai**
**Chen Xiliang**
**Cheng Ruiwen**
**Sofia Rodriguez**

Exhibition Design
**Cibic Workshop**

Graphic Design
**Hao**

Media
**Pang Weiran**
**Cao Yangqi**

Public Program
**Zou Zhuoer**
**Guan Yue**

特别鸣谢 With Thanks

Adriano Design · Agritube · Alessi · Alias · aMDL Circle · Archivio ADI Design Museum · Archivio Albe e Lica Steiner, Servizio Archivi Storici e Attività Museali, Politecnico di Milano, ACL · Archivio Carlo Orsi · Archivio Cassina · Archivio Centrale dello Stato · Archivio del Moderno, Fondo Marco Zanuso, Balerna · Archivio di Stato di Pescara - Fondo Corradino d'Ascanio · Archivio Riccardo Dalisi · Archivio Storico Solari di Udine · Archivio Storico Arflex · Archivio Storico ATM · Archivio Storico Fondazione Fiera Milano · Archivio Storico Olivetti · Arflex · Artemide · Felicia Arvid Jaeger · Associazione Archivio Storico Olivetti · Atelier Mendini · B&B Italia · Riccardo Blumer · Boffi S.p.A · Brionvega · Caimi Brevetti · Santi Caleca · Cassina · Centro Storico Fiat · Collezione Alessandro Pedretti design.Milano. · Collezione Paolo Rosselli · CSAC Centro Studi e Archivio della Comunicazione, Università di Parma · Dainese S.p.A · Danese Milano · Davide Groppi · Design Group Italia · D-Heart · Driade · Ducati · Editoriale *Domus* · Editoriale Lotus · Elica S.p.A · Fabita · Ferrari · Fiam Italia S.r.l · Flos · Fondazione Achille Castiglioni · Fondazione CSC – Archivio Nazionale Cinema Impresa (Ivrea) · Fondazione Franco Albini · Fondazione Jaqueline Vodoz e Bruno Danese, Milano · Fondazione Pirelli · Fondazione Vico Magistretti · FontanaArte · Foscarini Galleria Campari · Galleria Rossana Orlandi · Massimo Gardone · Francisco Gomez Paz · Grivel · Hackability · Giulio Iacchetti · IIT-INAIL Robotic Prosthetic Hand Hannes · Italdesign - Giugiaro · Isinnova S.r.l · Kartell Museo · Konstantin Grcic Design · Luceplan · Maurizio Corraini srl · Martinelli Luce S.p.A · Mauro Masera · Momodesign Museo Alessi · Museo Kartell · Oluce – Milano · Palomar S.r.l · Piaggio & C. S.p.A. · Pinarello S.r.l · Politecnico di Milano DAER - Dipartimento di Scienze e Tecnologie Aerospaziali Laboratorio Tecnologico – Silvio Ferragina, Cristian Ferretti · Matteo Ragni · Rapsel International S.r.l · Repower · Rexite · Marc Sadler · Sadler Associati Leo Torri · Studio Isao Hosoe Design · Studio Origoni Steiner · Tubes Radiatori · Università IUAV di Venezia – Archivio Progetti, Fondo Mauro Masera · Università IUAV di Venezia – Archivio Progetti, Fondo Giorgio Casali · Università di Padova · Gianluca Vassallo · Vibram · Ettore Vitale · Zanotta SpA – Italy · Federica Zanuso · Lorenza Zanuso · Susanna Zanuso

# 目录

# Contents

# 意大利设计：
# 文化的身份

安缇雅
意大利驻沪总领事

# Italian Design as Italian Cultural Identity

Tiziana D'Angelo
Consul General of Italy in Shanghai

展览"引领潮流七十年——意大利金圆规奖经典设计作品展"由米兰意大利工业设计协会博物馆、意大利特雷卡尼百科全书研究院和上海久事美术馆合作举办，获得了意大利驻沪总领事馆和意大利驻沪总领事馆文化处的大力支持。本次展览由梅特·加西亚·桑奇斯、乔万尼·科莫利奥、弗兰西斯卡·巴勒娜·阿里斯塔及凌敏共同担任策展人。展览特邀意大利知名设计团队西比克工作室完成展览空间设计。

意大利设计是极致卓越的代名词，从时尚到工业设计，它都完美体现了"意式天才"的精髓：一种将美感、创意、深厚的文化底蕴，与创造力、功能性以及先进的生产系统相结合的能力。

意大利设计不仅是风格选择，更是特定生活方式的精致体现，是传统与现代的细腻融合，是无与伦比的工艺水平与精益求精的细节理念相结合的产物。它是意大利文化认同的重要组成部分。事实上，制造业与文化的联结正是意大利制造的特色之一，它重申了我们的坚定信念，即文化不仅仅是一种"遗产"，更是一个强有力的概念，涵盖了我们的历史、我们的生活方式和思考未来的方式。

我们特别自豪能在上海久事美术馆举办这次展览。米兰和上海是世界上最具创造力和创新性的两座城市，也是联合国教科文组织"创意城市网络"的组成部分，这次展览进一步证实了两者之间由来已久的密切联系。

当我们在展览中肆意徜徉时，欣赏到的不仅仅是展品本身，更是一种文化叙事，是一个关于热情、创新和意大利设计永恒魅力的故事。

The exhibition "Compasso D'Oro Award: Seventy Years Leading Italian Design Trends" is the result of a joint effort between ADI Design Museum, the Institute of the Italian Encyclopedia Treccani and Jiushi Art Museum, with the support of the Consulate General of Italy, the Italian Institute of Culture, the curatorship of Maite Garcia Sanchis, Francesca Balena, and Giovanni Comoglio, with Ling Min, and the brilliant exhibition design of Cibic Workshop.

Italian design is synonymous with absolute excellence and, from fashion to industrial design, it represents perfectly the "Italian genius": the ability to combine an immense cultural heritage, beauty and creativity with innovation, functionality and advanced production systems.

Italian design is much more than just a stylistic choice; it is a sophisticated representation of a particular way of life, a fusion of tradition and modernity, the product of unparalleled craftsmanship and attention to detail. It is an essential part of Italian cultural identity. And, in fact, the link between manufacturing and culture is one of the elements that distinguish Made in Italy reaffirming our strong belief that culture is much more than "heritage", but it is a powerful concept that embraces our history, our lifestyle and the way we think about our future.

We are particularly proud to present this exhibition at the Jiushi Art Museum also because it represents a further confirmation of a strong link existing since a long time between two of the world's most creative and innovative cities, Milan and Shanghai, both part of the UNESCO "Creative Cities Network".

As we navigate the diverse landscapes of this exhibition, we are invited to appreciate not just objects, but a cultural narrative, a story of passion, innovation, and the timeless allure of Italian design.

# 上海 ADI 金圆规奖

达仁利  意大利驻沪总领事馆文化处处长

# ADI Compasso d'Oro in Shanghai

Francesco D'Arelli
Director of the Italian Cultural Institute
in Shanghai

"夫子之道，忠恕而已矣。"

——《论语 · 里仁第十五》

自古以来，没有规矩，不成方圆，线无所依，量无所凭，比例无所计。

圆规被古希腊人视为至雅至简的工具，在数学、天文、航海、测地、军事等领域应用广泛。其特质在设计领域尤受青睐，久已成为艺术与科学的标志性元素。

古罗马时期，圆规属于常见工具。在公元 1 世纪的庞贝古城遗址中就有圆规和其他精密工具的身影，是泥瓦工和木匠的常用工具，圆规同时也是各类装饰工程匠人的心头好。圆规的两脚间以三角形曲尺连接，顶端聚于一片折叶，便于调整开合。折叶可以安装固定件，锁定圆规角度。借助圆规可以画出完美的圆形，精准测量周长。

科技日益发展，各种专用圆规在文艺复兴时期逐一登场。莱奥纳尔多·达·芬奇（1452—1519）可称个中能手，研究制造了多种圆规，既有能够调节两脚开合的传统圆规，也有椭圆规和抛物线规，能够绘制圆锥曲线等更复杂的特殊图形。

达·芬奇对自然秉持着忠实无二的崇敬，以当时最纯粹的人文主义方式极致融合古典主义与文艺复兴，将人置于宇宙的中心，研究人体，确定完美的人体比例——由此诞生出以古罗马建筑家维特鲁威命名的传世名作《维特鲁威人》。维特鲁威在《建筑十书》中写道："若人仰卧，手脚伸展，以肚脐为圆心画圆，则手指和脚趾会恰好触及圆周线。"

圆规以纯金制成，贵金属的绝对价值也彰显着圆规的精到准确与无懈可击。黄金，在所有古文明中，都象征着太阳的光辉、神圣、圣洁、不朽。在跨越文明的文化交流中，规与尺简单结合而成的黄金分割唤醒人类对互惠这一哲学法则的记忆。

在曲与直、环与面、传统与文明的体系中，在知名作品与人物面前，正在上海举办的"引领潮流七十年——意大利金圆规奖经典设计作品展"自有其使命：在这里，记忆、理想、渴望、语言、产品、交流、碰撞、激荡。一如"两脚间以三角形曲尺相连"的圆规，在中国的天空下，中意文明相遇相知。唯愿这一盛典能在中国落地生根、花繁叶茂。正如品达罗斯的《颂歌》所吟唱，黄金是宙斯之子，却又脱离神的掌控；就让这把"金圆规"在上海驻足，画出生命的圆满。

"The way of the master consists of acting with the utmost loyalty and in not imposing on others what you do not want for yourself"
Lunyu, IV, 15

Since antiquity, the compass has served as a crucial design tool, enabling the creation of circles, the determination of angle bisectors, the drawing of perpendicular and parallel lines, the comparison of measurements, and the execution of proportional calculations.

The Greeks considered it the most elegant and simple construction tool, for it was adopted in the fields of mathematics, astronomy, navigation, topography, as well as military. Above all, it was considered particularly elegant in its simplicity due to its application in design. Since then, the compass has been widely used as a symbolic element of the arts and sciences.

In the ancient Roman era, the compass—*circinus*—was a common tool. It appeared alongside other precision tools in Pompeii's archaeological site from the first century CE. These were tools commonly used by workers such as bricklayers and carpenters, and it was employed in the projects of various decorative artisans. The compass consisted of two elements with a triangular section ending in a point and connected to a hinge which controlled its opening and closing. The hinge could be equipped with a fixing element to lock the compass at a certain angle. This tool was aimed at reproducing circumferences and reporting measurements with high precision.

Successively, following the development of technology in the field, many special compasses were created during the Renaissance. Leonardo da Vinci was evidently no stranger to the design and use of such an instrument; he studied and worked on the development of various models, both for traditional use with an adjustable opening, and for special and complex uses, e.g., for drawing conical curves, such as ellipses (the so-called ellipsograph) or parables (parabolic compass).

Condensing and integrating Classicism and Renaissance in the purest humanist expression of his time, Leonardo—a profound observer of Nature—put Man at the center of the universe, studied the human body, and set its ideal proportions. This work resulted in the *Vitruvian Man*, a work inspired by the writings of the ancient Roman architect Vitruvius and in which we read "… if a man lies on his back, with his hands and feet extended, and places the end of a compass on his navel, the fingers and toes will touch the circumference of the circle that we thus draw."

The precision and perfection of the compass as a tool is here combined with the absolute value of the metal of which it is made — gold. In the transversality of ancient civilisations, gold—associated with yellow—is a symbol of solar splendor, sacrality, divine filiation, and incorruptibility. In the context of cultural dialogue between civilizations, the golden section, simply constructed with a compass and straightedge, recalls to a philosophical rule of reciprocity among humans.

It is no coincidence at all that in this system of lines and curves, circles and circumferences, traditions and civilisations, as well as in eminent figures, the *ADI Compasso d'Oro: Seventy Years Leading Italian Design Trend* exhibition now taking place in the megalopolis Shanghai, has an innate vocation: to be a place for the exchange of memories, aspirations, desires, languages, and goods. As the compass, "two elements with a triangular section," we see the dichotomy of Sino-Italian cultural dialogue visualised under the heaven of China. It is my wish for this event to land in this place so we may see it as an outstanding sprout, and to encourage this essential sap to grow, develop, and procreate. As we read in Pindar's *Odes*, "I remember that gold is the son of Zeus and it flows from the god's own might;" let this Golden Compass set its feet in Shanghai and design remarkable circles of life.

# 意式卓越设计史

马西莫 · 布莱
意大利特雷卡尼百科全书研究院总经理

# A History of Italian Excellence

Massimo Bray
Director General of the Institute of the
Italian Encyclopedia Treccani

"设计是指为家用或工作相关用途的物品、产品、工具制定规划方案的活动。设计产品可手工打造，也可大规模生产，但始终兼具技术性与美观性（每个设计都是一个完整的项目）。"以上是意大利特雷卡尼百科全书研究院出版的2006年《意大利百科全书》附录中，由意大利知名建筑师、设计师安德烈亚·布兰齐在"设计"这一词条下给出的定义，还介绍了意大利和其他国家的设计简史。

毋庸置疑，意大利长期引领着美学产品，特别是时尚和设计领域的国际潮流。20世纪50年代起，"意大利制造"一直对世界各地眼光独到的消费者有着无法抗拒的吸引力，是品质和品位的代名词，最终成为意大利的一大经济支柱。意大利设计在工业和文化两个层面都与众不同，最终在世界范围内脱颖而出，取得了公认的成就。一如布兰齐所说："意大利的设计别具匠心，把瑕疵和弱点变成新设计的优势，化腐朽为神奇。"

此次展览由上海久事美术馆与米兰ADI设计博物馆联袂呈现，讲述意式卓越设计的迷人历史，展示经由蜚声全球的金圆规奖筛选而来的标志性作品、影像、资料、沉浸式工具。

意大利工业设计协会自创立以来，始终关注设计的内在价值，"意式设计"更是重中之重。这种考量兼顾设计理论和实用性，最终凝结为"金圆规奖"。金圆规奖极具代表意义，更是体现ADI设计原则和设计菁华的全球设计领域标杆。70年匆匆而过，金圆规奖至今仍是全球设计顶级奖项，肩负机构类奖项的声望。尽管不敢夸口参评作品将全球优秀设计尽数网罗，但金圆规奖无疑是广阔设计领域内的权威大奖。

设计立足于坚固的科学技术基础，是一门跨学科艺术，弥合了领域间的差异。当天差地别的事物经由设计浑然一体，设计也就凭借这一独特的能力，既是每日生活不可或缺的日常，也是瞬息万变世界中复杂问题的解决之道。正因如此，意大利特雷卡尼百科全书研究院一直积极支持设计艺术，借由相互理解、共同努力，寻求、推动卓越文化，应对意大利和全球问题。

"The term design refers to the activity of planning objects, products, or tools, whether for domestic or work-related use. These items can be crafted manually or produced industrially, where technical aspects coexist with aesthetic considerations (design = project)." Renowned Italian architect and designer, Andrea Branzi, provided this definition and a brief history of design in Italy and abroad in his *2006 Italian Encyclopedia Appendix* entry, published by Treccani.

Italy has undoubtedly held a global leadership position in the market of aesthetic products, particularly in fashion and design, for an extensive period of time. Since the 1950s, the "Made in Italy" label has been a compelling attraction for discerning consumers worldwide, guaranteeing a high level of technical quality coupled with sophisticated expressive capacity. This has ultimately become a cornerstone of the national economy. The globally acknowledged success stems from distinctive elements that set the Italian system apart, both industrially and culturally. As noted by Branzi, "Italian design has uniquely interpreted these diversities, transforming some of its flaws or weaknesses into positive opportunities for new strategies."

This exhibition, organized by Jiushi Art Museum, Shanghai, and ADI Design Museum, Milan, recounts a captivating history of Italian excellence. It showcases iconic objects, images, documents, and immersive tools, viewed through the lens of the prestigious Compasso d'Oro Award.

Since its establishment, the Association for Industrial Design (ADI) has consistently contemplated the intrinsic values of design, with a particular focus on the esteemed Made in Italy design. This reflection, encompassing both theoretical and practical aspects, takes tangible form through the ADI Compasso d'Oro award. Beyond its symbolic significance, this award serves as a prominent global reference point, encapsulating the essence of design according to the principles of the ADI. Now celebrating its seventieth year, the Compasso d'Oro remains the foremost global design accolade, representing a prestigious institutional honor. Though not exhaustive, it offers a plausible and valid measure of qualitative value within the expansive design landscape.

Distinguished by its solid technical and scientific foundations, design acts as a conduit for interdisciplinary knowledge, bridging gaps among diverse disciplines. Possessing a unique ability to synthesise widely separated experiences and situations, it offers comprehensive solutions to everyday life and addresses the profound questions posed by our ever-changing world. It is for this very reason that the Institute of the Italian Encyclopedia has supported this project, representing an extraordinary opportunity to engage in an enterprise of mutual understanding and joint effort towards the recognition and enhancement of cultural excellence, responding to both local and global issues.

# 设计之必要

卢恰诺 · 加林贝蒂　ADI 主席

# Necessary Design

Luciano Galimberti
ADI President

正如马西莫·卡奇亚里的"必然的天使"一样，设计也与我们时时刻刻形影相随：从晨起到暮落，甚至一同入梦。我们享用早餐，我们洗漱，我们穿衣，我们出行，我们探索，我们工作，我们享乐，我们尽情去爱……

行之所至，目之所及，我们日常生活中的种种空间、物体、服务与界面几乎都离不开行之有效的设计。因此，设计这一学科肩负重责大任。

毋庸置疑，无论是在当下还是未来，设计的角色作用往往取决于彼此关联甚微的学科、知识、兴趣之间的纷繁关系与多样互动。虽然深深植根于科技的沃土，但设计与其他科学类学科的不同之处在于，它能够促进学科交融，在割裂的体验与情境间寻求平衡。否则，面对全球化所提出的宏大命题，我们将无法给出完整的答案。

ADI 自创立以来，一直不断"延长当下"，深切领悟每时每刻，从而借此反思设计，尤其是意式设计究竟代表怎样的价值体系。在思考的基础上，我们也付诸行动——ADI 金圆规奖无疑最为清晰地阐释了何谓我们所说的"设计"，同时也已成为全球设计界亮眼的标杆。

走过 70 年漫漫长路，金圆规奖始终是设计界最炙手可热的奖项：这不是我的一家之言，而是国际设计界的一致认可。该奖项长期蜚声国际，一是因其极为苛刻：70 年来仅颁发了 350 余项；二是因其评选过程科学，由 150 名多学科专家组成的常任观察委员会负责；但最重要的是，金圆规奖作为机构奖，绝非像许多其他知名奖项一样秉持商业逻辑。

ADI 为自己的选择颇感自豪——纵使评选过程往往艰辛，我们始终坚持自身对设计的理解。当然，仅凭一个奖项，并不能全面展现如设计体系一般错综复杂的系统，但历届获奖作品作为可靠的指标，能够帮助我们体悟设计的本质价值所在。

70 年来，共计二十七届金圆规奖为我们留下了宝贵的财富。ADI 通过基金会将图像、经验与作品收藏至米兰的 ADI 设计博物馆，展出推广。ADI 设计博物馆既是研究型博物馆，也是当代的创新平台，能够让人们置身全球复杂的设计产业链中，发挥奇思妙想、交流体会，培育梦想。

我谨在此衷心感谢意大利驻沪总领事馆及文化处和上海久事文化传播有限公司为在上海举办此次重要国际盛事所提供的大力支持。我坚信这将不断激发我们的灵感与思路，面对如今日益互联互通的世界，共同思考并携手向前。

Like Massimo Cacciari's "necessary angel", design is with us through each of the twenty-four hours in our day: from waking up to going back to bed again, not leaving us even during the night. We have breakfast, we wash ourselves, we get dressed, we move from one place to another, we discover things, we work, we have fun, we make love...

Spaces, objects, services and interfaces; practically every step of our everyday life is accompanied and supported by design and its operational methodology; as such it is a discipline with considerable responsibilities.

Today, and certainly tomorrow, the role of design will have to be defined within an increasingly wide-ranging panorama of relationships and interactions involving disciplines, knowledge and interests that are often very distant from each other. Design is a discipline with solid technical-scientific foundations, which differs from all other scientific disciplines in its ability to facilitate interdisciplinary knowledge, balancing widely separated experiences and situations, which on their own would only give partial answers to the big questions that the globalized world is asking us.

Since its foundation, ADI has reflected through a sort of "long now" on the system of values to which design and in particular Made in Italy design refer. This reflection combines both thinking and doing and our ADI Compasso d'Oro award is undoubtedly the most evident embodiment of what the Association means by the term design as well as being a clearly visible benchmark for the rest of the world.

Now in its seventieth year, the Compasso d'Oro remains the world's most coveted design award: not my words, rather the unanimous acknowledgement of the world of international design. The award has always stood out for the parsimony with which it has been given; the fact remains that in seventy years of history just over 350 awards have been conferred. It is also an award that stands out on account of the scientific selection process that involves a permanent observatory made up of 150 multidisciplinary experts, but above all it is an institutional award, far from the commercial logic of many other prestigious forms of recognition.

ADI is proud of its choices, these are often tiring and uncomfortable, but always consistent with its way of understanding design. An award certainly cannot be exhaustive in the representation of a system – complex as it is – such as that of design, but the snapshot that each edition gives us is a plausible and certainly useful measure of qualitative value.

In its seventy years of history twenty-seven editions have created a legacy of images, experiences and products that through its Foundation, ADI manages to preserve and promote in the ADI Design Museum in Milan. This is a museum for research and a contemporary and innovative platform, capable of developing all the ideas, experiences and dreams related to the complex design chain on a global scale.

I would like to thank Consolato Generale d'Italia in Shanghai, Istituto Italiano di Cultura in Shanghai, Shanghai JUCE Culture and Media Co., Ldt., for their support in the establishment of this important international event in Shanghai. I am sure that it will be one which will help us to constantly find new ideas for common reflection and working together in an increasingly interconnected world.

# 设计之美  文明之桥

上海久事美术馆

# Design as a Bridge
# Between Civilizations

Shanghai Jiushi Art Museum

自20世纪中叶以来，意大利设计在全球范围内一直引领着美学与实用性的融合潮流。这一地位的奠定，离不开其国家对设计价值的深度挖掘与广泛认可。而今，这一传统在新时代得以延续，特别是在与中国的交流与合作中。

意大利的"金圆规奖"是设立时间最早、历时最悠久的设计奖项，从诞生至今它见证了意大利70年来的设计风潮与变迁。这一奖项不单是对设计行业从业者所产出贡献的认可，更是对意大利设计理念和工艺的肯定。每一件获奖作品从不同角度出发，展现了其技术、美学和实用性，让观众能够从多维度领略意大利设计的独特魅力。

2023年伊始，上海久事美术馆有幸与意大利工业设计协会结缘。在双方的共同努力下，我们将意大利设计的瑰宝——"金圆规奖"引入中国。在这个名为"引领潮流七十年——意大利金圆规经典设计作品展"的展览中，我们首次系统、完整地将金圆规奖的经典作品、文献资料、历史影像等介绍给了中国观众，完美呈现了这一享誉全球的设计奖项70年的辉煌历程。

从传统的工艺品、现代的科技产品到公益项目，我们可以看到"金圆规奖"的逐步成熟，它的多元化和包容性，以及对社会的关注和反思。中国和意大利两国在设计领域都有着悠久的历史和独特的传统。中国设计讲究和谐、自然和实用性，而意大利设计则注重创意、美学和工艺。虽然两国在设计理念上有所不同，但正因为交流与合作，才能在设计交融中带来更多的创新和可能性。我们也期望看到中、意两国在设计产业方面的合作与发展。在全球化日益加速的今天，设计师们面临着共同的挑战和机遇。两国都有着各自的优势和特点，如果能够将这些优势和特点结合起来，以满足日益多样化的市场需求，必将产生更大的经济和社会效益。

如今随着全球互联的深入发展，设计已不再局限于某一国或某一文化。它成为了一种跨文化、跨领域的交流方式。而意大利设计正是这一趋势的引领者，它以其独特的审美和工艺，影响着全球的设计潮流。此次展览是上海久事美术馆在艺术展览方面向设计领域迈出的第一步。我们希望通过这样的展览，让更多的观众了解和欣赏到来自全世界的优秀设计。同时，我们也期待借由此次展览，中、意两国在设计领域的合作更加紧密，共同推动全球设计的发展与创新。

最后感谢参与本次展览的各方，意大利共和国驻上海总领事馆及文化处、意大利工业设计协会、米兰ADI设计博物馆、意大利特雷卡尼百科全书研究院以及凌敏女士。我们期待在这些合作伙伴的共同努力下，我们将用金圆规画出更多完美的黄金螺线。

Since the mid-20th century, Italian design has been leading the global trend of integrating aesthetics with functionality. Such a leadership stems from nationwide recognition and in-depth exploration of the intrinsic value in this field. In the new era, this tradition endures, especially in exchanges and cooperation with China.

Compasso d'Oro, as a design award with the longest history, bears witness to the evolving trend of Italian design in the past seven decades. It not only honours the contributions made by producer and designer community, but also recognises the concept and craftsmanship emblematic of Italian design. Each award-winning design, by demonstrating its combination of aesthetics, technology and practical solution from various perspectives, engages the audience with a multifaceted experience of the unique Italian design.

In early 2023, thanks to the collective effort of Shanghai Jiushi Art Museum and Association for Industrial Design (ADI), Compasso d'Oro, the gem of Italian design, was introduced to China. For the first time, the exhibition *Compasso d'Oro Award: Seventy Years of Leading Italian Design Trends* presents Chinese audiences with a panorama of this world-renowned award. It features classic pieces, archives, and historical images, fully capturing the glorious journey throughout the years.

From traditional handicrafts to modern high-tech products, and to public welfare projects, Compasso d'Oro is evolving towards greater diversity and inclusiveness, while following closely and reflecting on social development. Both Chinese and Italian designs boast a long history and distinct tradition. While the former prioritizes practicality in harmony with nature, the latter focuses more on creativity, aesthetics and crafts. Despite such differences, communication and cooperation will inspire more innovative possibilities. Bilateral collaboration in the design industry is thus highly welcome. Amidst accelerating globalisation, all designers face common challenges and opportunities. If respective strengths of both nations can be pooled to cater to the needs of an increasingly diversified market, economic and social benefit will surely amplify.

As the world is becoming increasingly interconnected, design has transcended geographical and cultural borders, becoming a means of communication. Italian design, with its unique taste and crafts, continues to be a leading influence in the global design arena. This exhibition marks our Museum's first foray into the art of design. We wish initiatives like this will make excellent designs from all countries more accessible, and enhance bilateral cooperation in design, thus driving innovation in the global design community.

In conclusion, we'd like to extend our sincere appreciation to Consulate General of the Republic of Italy in Shanghai and Italian Cultural Institute in Shanghai, ADI, ADI Design Museum in Milan, Treccani, and Ms. LING Min. With the joint effort of all the partners, we are eager to draw more perfect golden spirals together with Compasso d'Oro.

# 金圆规奖：
# 引领潮流 70 年

# Compasso d'Oro Award:

# Seventy Years Leading
# Italian Design Trends

金圆规奖为彰显优秀的工业设计而设，涵盖设计类型多样，从产品设计到当代诸多模糊的表达；表彰具有设计远见的项目、人物、制造商，正是他们不断拓展学科边界，用巧妙的方式诠释所处的时代。

确切地说，金圆规奖旨在发现设计领域中的杰出作品：早在 1954 年创立之时，金圆规奖就开创性地认识到，生产不仅是一种经济现象，更包含美学、文化与习俗。秉承这一精神，本次展览回溯奖项的历史，探寻金圆规奖在历史背景下和一系列社会文化风潮中的重要意义——这些风潮不仅界定了意式设计，也是当代国际工业设计的叙事主线。

因此，本次展览以主题式展陈为指导思路，在由一百多个获奖项目接续而成的时间线上，点缀若干主题空间，从金圆规奖的审辨研究视角，彰显在当代具有重要意义的设计主题。

本次展览挖掘金圆规奖 70 年的深厚积淀，展品全部来自奖项创设以来的独特收藏，包括实物、文件与相关叙述。在历史藏品的基础之上，藏品档案仍生机勃勃，不断壮大。此外，展览涉及的各家机构、品牌与档案，完美呈现出意式设计文化的博大精深、错综复杂，交织而成广阔鲜活的网络，勾勒出金圆规奖注重合作的本质内涵。

The Compasso d'Oro award celebrates industrial design at large, in its multiple forms – from product design to contemporary intangible expressions – and recognises projects, people and manufacturers with a design-driven vision, that push the disciplinary boundaries ever farther and provide a clever interpretation of their time.

The mission of the Compasso d'Oro award is in fact to identify outstanding realities in the realm of design: at its origin in 1954, the award was the first official recognition addressing the problem of production as a fact not only of economics but also of aesthetics, culture and custom. Following this spirit, the exhibition is staged as a journey which presents the history and relevance of the award in close relation to its historical context and to a series of social and cultural trends that, on top of defining Italian design, have become fundamental narrative threads of contemporary international industrial design.

This thematic approach is therefore the guiding principle in the exhibition experience, where a chronological landscape, of over one hundred awarded projects, is punctuated by a series of spaces providing thematic insights on contemporary relevant subjects, stressed by the critical research and investigation of Compasso d'Oro.

The source for this seven-decade-deep exploration is the unparalleled collection of objects, documents, and testimonies built by Compasso d'Oro over its lifespan. It is articulated in a living and ever-growing archive and a historic collection, supported in the development of this exhibition by another entity that speaks clearly about the collaborative nature of the award: an extensive and vibrant network of interconnected institutions, brands, and archives that directly express the breadth and complexity of Italian design as a cultural landscape.

# 设计一个奖项

# Designing an Award

金圆规奖的标志由平面设计师阿尔贝·斯坦纳于 1953 年设计，其灵感来自画家阿达尔伯特·戈林格于 1893 年发明的黄金比例除法器。授予金圆规奖获奖者的金圆规是斯坦纳符号的具体化，由马尔科·扎努索和阿尔贝托·罗塞利于 1954 年设计。

The symbol of the Compasso d'Oro (Golden Compass) was designed by graphic designer Albe Steiner in 1953, inspired by the Golden Ratio Divider invented by the painter Adalbert Goeringer in 1893, a tool used to determine whether two measurements are related through the Golden Ratio. The golden compass awarded to the winners of the Compasso d'Oro Award is the physical translation of Steiner's symbol and was designed by Marco Zanuso and Alberto Rosselli in 1954.

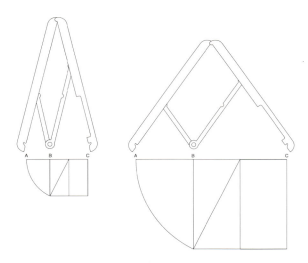

黄金矩形与金圆规。

The construction of the golden rectangle with the golden compass.

鹦鹉螺，自然界中最完美的金色螺旋之一；金色分割线，Adalbert Goeringer，1893 年；金圆规，Albe Steiner 的品牌设计，1954 年；金圆规奖，Marzo Zanuso 和 Alberto Rosselli 设计。

Nautilus; Golden divider, Adalbert Goeringer, 1893; Compasso d'Oro, brand design by Albe Steiner, 1954; Compasso d'Oro Award, designed by Marzo Zanuso and Alberto Rosselli.

# 黄金分割：
## 自然神工与
## 人工巧技的碰撞

# Golden Ratio:

# The Encounter between

# Nature and Artifice

黄金比例是自然界中十分常见的数学比例，见于各种图案、排列和植物构造之中，是数学家自古以来研究的课题。全世界的重要艺术品和建筑是人类共有的美学遗产，文化影响深远，于其中也常有黄金比例（又称"神圣比例"）的身影。

黄金比例是约为 1.618 的无理数，既非整数，亦非分数。因此，通常使用无止境的对数螺旋图形来表示，即处在一系列长宽比为黄金比例的黄金矩形之内的黄金螺旋。要画出黄金螺旋，需要借助与圆规相似的黄金比例测绘工具——黄金分线规。

将黄金螺旋叠加于自然界与人造物之上，螺旋的几何及数学特性显露无遗，这个图形反复出现在看似毫无关联的自然和人文场景里，使人们从美学和自然文化遗产的新视角理解设计。

The Golden Ratio is a mathematical proportion frequently found in nature – in patterns, arrangements, and other parts of vegetation – that has been studied by mathematicians since antiquity. Also known as the Divine Proportion, this ratio is also found in key artistic and architectural expressions throughout the world, which have defined a common aesthetic heritage for humankind and continue to exert a significant cultural influence.

The ratio is an approximation of the irrational number 1.618…; it can't be described as either a whole number or a fraction. Therefore, it is usually represented graphically with an infinite logarithmic spiral, known as the golden spiral, contained inside a sequence of rectangles that relate to each other through the Golden Ratio. This spiral can be drawn with the help of the golden divider, a tool for measuring and constructing this ratio, similar to a compass.

When superimposing the golden spiral onto natural and artificial elements, its geometric and mathematical properties, as well as its reoccurrence in a number of apparently unconnected natural and cultural contexts, become visible to the eye. This opens a new way of understanding design in relation to aesthetics and our natural and cultural heritage.

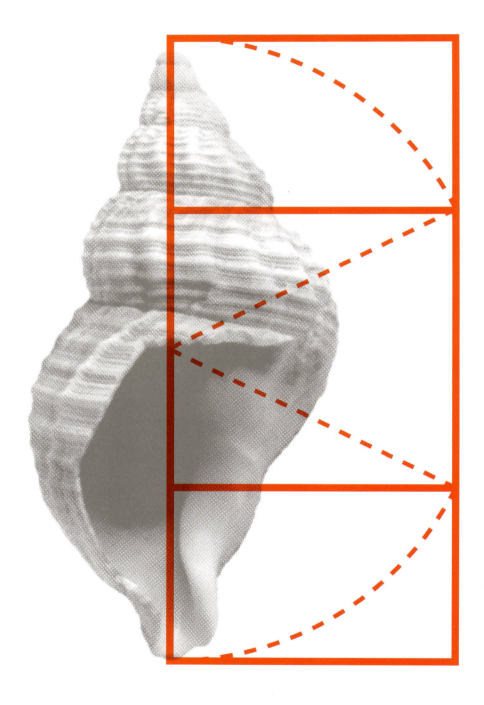

贝壳。
Seashell.

海星。
Starfish.

© 阿尔多 · 蒙图《关于黄金分割和五边形的注释和说明》
© Aldo Montù, Appunti ed annotazioni su sezione aurea e
forme pentagonali

贝壳耧斗菜花。
Aquilegia flower.

雪晶。
Snow crystal.

© 阿尔多 · 蒙图《关于黄金分割和五边形的注释和说明》
© Aldo Montù, Appunti ed annotazioni su sezione aurea
e forme pentagonali

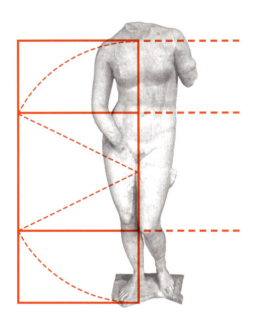

阿佛洛狄忒·科尼狄亚，公元前 2 世纪，
普拉克西特列斯作。
Afrodite Cnidia, 2nd century B.C.,
Prassitele.

斯特拉迪瓦里"布伦特女士"小提琴，
1721 年，安东尼奥·斯特拉迪瓦里制。
Stradivari Lady Blunt violin, 1721,
Antonio Stradivari.

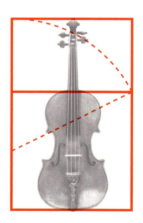

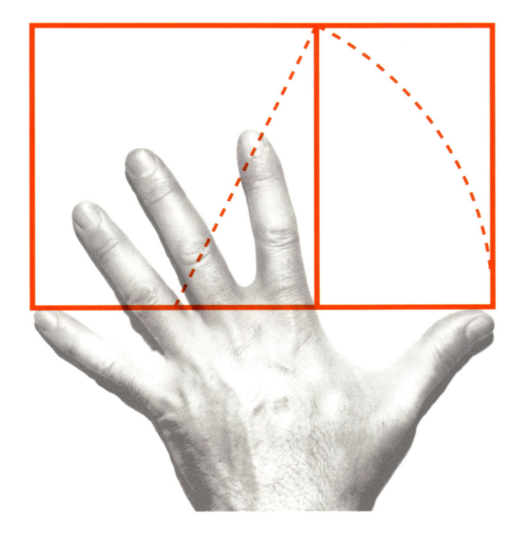

手。
Hand.

© 阿尔多 · 蒙图《关于黄金分割和五边形的注释和说明》
© Aldo Montù, Appunti ed annotazioni su sezione aurea
e forme pentagonali

《神奈川冲浪里》，1830 年，葛饰北斋绘。
The Great Wave, 1830, Katsushika Hokusa.

CC0 公共领域专用
CC0 Public Domain Dedication

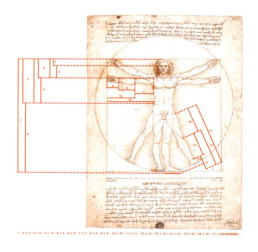

《维特鲁威人》，约 1490 年，莱奥纳尔多 · 达 · 芬奇作。
Vitruvian Man, ca.1490, Leonardo da Vinci.

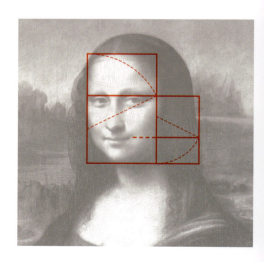

《蒙娜丽莎》，1503 年，莱奥纳尔多 · 达 · 芬奇作。
Mona Lisa, 1503, Leonardo da Vinci.

公共领域专用
Public Domain Dedication

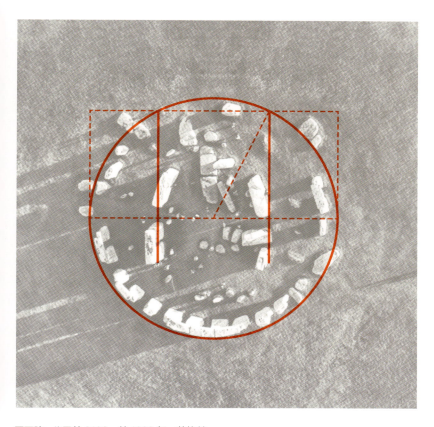

巨石阵，公元前 3100—前 1600 年，英格兰。
Stonehenge, between 3100 and 1600 B.C., England.

© iStock.com/Michael Chapman

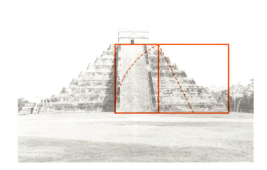

库库尔坎金字塔，公元 800—1200 年，墨西哥。
Kukulcán temple, between 800 and 1200 A.D., Mexico.

© iStock.com/Csondy

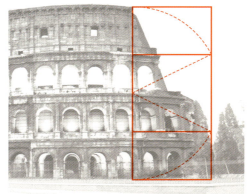

圆形竞技场，公元 72 年，意大利。
72 A.D., Italy.

© iStock.com/spooh

# 金圆规奖
# 探秘之旅

# A Journey through the
# Compasso d'Oro Award

金圆规奖"开创性地认识到，生产不仅是一种经济现象，而且还涉及美学、文化与习俗问题"。
*Stile Industria* 杂志，1954 年第 1 期。

金圆规奖于 1954 年由著名的文艺复兴百货公司创立，自 1959 年起与 ADI（意大利工业设计协会，成立于 1956 年）合作筹办；1967 年开始，ADI 完全接管奖项的管理与策展工作。最初规定"所有意大利工业与手工艺设计公司以及模型制作者"均可报名，角逐大奖。

金圆规奖代表意大利最顶尖的设计文化，由战后的设计界大咖携手开创，包括 *Domus* 杂志创始人、建筑师、设计师吉奥 · 蓬蒂，*Stile Industria* 杂志创始人、总监阿尔贝托 · 罗塞利，文艺复兴百货公司艺术总监、平面设计师阿尔贝 · 斯坦纳。

随着时间推移，历届金圆规奖的举办日益清晰地呈现出这一奖项的发展脉络，从最初的工业设计推动者逐渐成长为意大利与国际设计体系的瞭望台与驱动力。

早在 20 世纪 60 年代，金圆规奖就开始日渐演进：获奖主题和类别先是扩展至新的设计领域；在项目奖依然保留的情况下，大奖逐渐合并为颁发给个人与公司的奖项，于 1994 年融合为现在的金圆规终身成就奖（从 2020 年开始，终身成就奖也颁发给长盛不衰和经典标志性的产品）。1998 年，意大利常任设计观察组成立，确认了 ADI 在定义当代设计版图方面发挥的作用，观察组所挑选的优秀设计作品又汇集成为《ADI 设计索引》，最终获奖作品就从中诞生。

这一奖项的社会功能也在与时俱进：多年来，设计作品及相关文件已经形成了规模庞大的收藏。2021 年，这些收藏终于得以在米兰的 ADI 设计博物馆安家落户。博物馆传达了"金圆规奖"的使命——在设计界与用户、从业者和普通民众所组成的日益壮大的参观者群体之间，激发文化层面上的思辨与碰撞。

The Compasso d'Oro Award was "the first event to recognise and address the problem of production, not only as an economic fact but also as a matter of aesthetics, culture, and customs".
*Stile Industria* n. 1, 1954.

Established in 1954 by La Rinascente, a prominent department store, since 1959 the award was organized in collaboration with ADI, Associazione per il Disegno Industriale (Italian Association for Industrial Design, founded in 1956), which has been the sole responsible for its management and curation since 1967 up to the present day. According to the initial regulations, "all Italian industrial and artisan companies and the creators of the models" could compete for the award.

The award is an expression of the best Italian design culture. Some of the most influential figures in the post-war era participated in its conception and development. These included architect and designer Gio Ponti, founder of *Domus* magazine; Alberto Rosselli, creator and director of *Stile Industria* magazine; and graphic designer Albe Steiner, art director of La Rinascente.

The evolutionary nature of the Compasso d'Oro award has become more and more clearly delineated as the years and editions progressed, expanding its role from promoter of industrial design to observatory and driving force of the design system within the Italian and international context.

The awards started evolving as early as the 1960s: the subjects and categories awarded first of all expanded to new unexplored areas of the design realm; while the awards to projects remained, the Grand Prizes went merging into awards given to people and companies, crystallizing then since 1994 in the Compasso d'Oro alla Carriera (Career Award, from 2020 recognising the career of products as well, for longsellers and icons). In 1998 the Permanent Observatory of Italian Design was born, confirming ADI's role in defining an atlas of contemporary design, translated into the ADI Design Index system that has also become the  collector of selected Compasso nominations.

The social role of the prize has also been evolving: over the years it has built a substantial collection of objects and documents, which in 2021 have found a physical space, The ADI Design Museum in Milan, an expression of Compasso d'Oro's vocation to generate culture, debate and encounter between the world of design and an increasingly wide audience of users, practitioners and citizens at large.

在 70 年的发展历程中，金圆规奖持之以恒，表彰设计领域内的杰出成就。最初，奖项仅由评委会予以颁发；但之后，奖项体系日益复杂，设有多个主题评审委员会，并进行层层筛选评估。

1955 年，意大利金圆规奖和国际金圆规奖正式设立，旨在表彰突出的个人或组织。这两个奖项后来合并为金圆规终身成就奖。历届获奖作品和项目均收录在当年出版的合集目录中。

Over its seventy years of history, the Compasso d'Oro award has recognised outstanding achievements in design. Initially, this recognition was granted solely by its juries, but later it evolved into a more complex system involving thematic commissions and various levels of assessment.

In 1955, the Grand National Prize and the Grand International Prize – later merged into the current Compasso d'Oro Career Award – were established, to award individuals or organizations. Each edition is accompanied by the publication of a catalogue collecting the projects that have received the award.

1953 年，文艺复兴百货公司提议举办题为"产品美学"的展览，即金圆规奖的前身。这是首次由一家百货公司来解决产品的"广义尊严"这个"社会问题"。展览宣传册中这样写道："无论是一把刀、一条裙子，还是一只玻璃杯、一个熨斗，产品的尊严与价值不仅体现在价格与功能上，也关乎外观形态……"

In 1953, La Rinascente proposed the exhibition *L'estetica nel prodotto* (Aesthetics in the Product), which served as a precursor to the Compasso d'Oro Award. For the first time, a department store addressed the "social problem" of the "greater dignity" of the product. The presentation brochure reads, "A knife, a dress, a glass, an iron, have their dignity not only in price or function but also in form (...)."

▲ 1954 年首届文艺复兴百货公司金圆规奖颁奖仪式上，评委会成员、设计师吉奥·蓬蒂和阿尔多·博莱蒂。
1954 Award ceremony of the 1st La RInascente Compasso d'Oro award with designers Gio Ponti and Aldo Borletti, members of the jury.

ADI 设计博物馆档案
Archivio ADI Design Museum

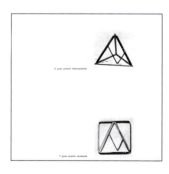

▲ 1955 年第二届文艺复兴百货公司金圆规奖目录中的国际金圆规奖和意大利金圆规奖。
1955 Awards of the Grand International Prize and the Grand National Prize category as presented in the catalogue of the 2nd La Rinascente Compasso d'Oro.

ADI 设计博物馆档案
Archivio ADI Design Museum

▲ 1955 年第二届文艺复兴百货公司金圆规奖目录中不同类别的奖项：金圆规授予获奖制造商；银盘授予获奖设计师；荣誉证书授予入围制造商。
1955 Awards of the various categories as presented in the catalogue of the 2nd La Rinascente Compasso d'Oro award: a golden compass for the winning producers, a silver plate for the winning designers and an honorary distinction diploma for the producers admitted to the competition.

ADI 设计博物馆档案
Archivio ADI Design Museum

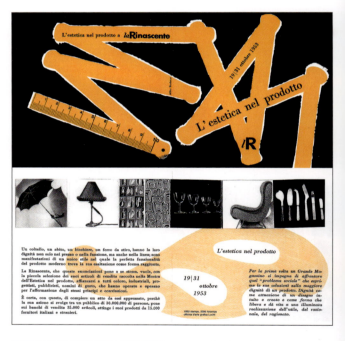

▲ 1953 年米兰文艺复兴百货公司的产品美学展览，由卡洛·帕加尼、布鲁诺·穆纳里、阿尔贝托·罗塞利策展，阿尔贝·斯坦纳负责平面设计。
1953 Exhibition *L'estetica nel prodotto* (Aesthetics in the Product) curated by Carlo Pagani, Bruno Munari and Alberto Rosselli, graphic design by Albe Steiner, at La Rinascente, Milan.

▲ 1953 年阿尔贝·斯坦纳为米兰文艺复兴百货公司的产品美学展览设计的宣传册。
1953 Brochure by Albe Steiner of the exhibition *L'estetica nel prodotto* (Aesthetics in the Product) at La Rinascente, Milan.

Origoni Steiner 联合建筑师工作室
Studio Origoni Steiner

米兰理工大学历史档案与博物馆活动服务，阿尔贝与莉卡·斯坦纳工作室
Archivio Albe e Lica Steiner, Servizio Archivi Storici e Attività Museali, Politecnico di Milano, ACL

1954 年，首届金圆规奖评奖活动举行，当时该奖项仍叫做"文艺复兴百货公司金圆规奖"。获奖作品在第十届米兰三年展中展出。三年展是一家在意大利文化讨论中颇为活跃的重要机构。直至 1957 年，前四届金圆规奖的举办与管理工作均由文艺复兴百货公司负责。

In 1954, the first edition of the prize, then called La Rinascente Compasso d'Oro, was held. The exhibition of the award-winning products was hosted within the 10th exhibition of Triennale di Milano, a significant institution and one of the most active in the Italian cultural debate. La Rinascente produced and managed the first four editions until 1957.

二战后的意大利诞生了工业设计文化。*Stile Industria* 杂志由阿尔贝托·罗塞利在 1954 年创刊，是意大利第一份专门探讨工业设计的刊物，在激发讨论、塑造品味方面发挥了重大作用。

In post-war Italy, the culture of industrial design was born. The magazine *Stile Industria*, founded in 1954 by Alberto Rosselli, was the first in Italy to be entirely dedicated to this subject and played a fundamental role in the development of the debate and the shaping of taste.

▲ *Stile Industria* 杂志，1954 年 6 月，第 1 期。
*Stile Industria* n. 1, June 1954.

*Domus* 杂志档案
© Editoriale *Domus* 联合股份有限公司
Archivio *Domus* © Editoriale *Domus* Spa

▲ *Stile Industria* 杂志，1961 年 3 月，第 31 期。
*Stile Industria* n. 31, March 1961.

*Domus* 杂志档案
© Editoriale *Domus* 联合股份有限公司
Archivio *Domus* © Editoriale *Domus* Spa

▲ 1954 年首届文艺复兴百货公司金圆规奖展，由卡洛·帕加尼和吉安·卡洛·奥尔泰利设计，第十届米兰三年展。
1954 Exhibition of the 1st La RInascente Compasso d'Oro award, design by Carlo Pagani and Gian Carlo Ortelli, X Triennale di Milano.

ADI 设计博物馆档案
Archivio ADI Design Museum

自创立以来，金圆规奖的重要意义还在于向日益庞大的公众群体进行大力宣传。精选项目在各类机构和重要的文化空间展出，展陈均由意式设计大咖构思设计。

Since its foundation, the significance of the award also lies in its role of diffusion to an ever-widening public. Institutions and significant cultural spaces host exhibitions of the selected projects, staged in settings conceived by important figures in Italian design.

金圆规奖的国际性从设立之初即已凸显，1957 年参加纽约世界博览会这样的盛事更是其高光时刻。公众借此机会，能够亲身感受意式设计文化的独特魅力。

The international scope of the Compasso d'Oro award, present from the very beginning, was highlighted by its participation in events like the New York World's Fair in 1957. The public is directly engaged in a genuine encounter with the culture of Italian design.

▲ 1955 年马塞尔·布劳耶国际金圆规奖作品展，米兰塞尔贝洛尼宫新闻俱乐部。
1955 Exhibition for the Grand International Prize awarded to Marcel Breuer, Circolo della Stampa, Palazzo Serbelloni, Milan.

ADI 设计博物馆档案
Archivio ADI Design Museum

▲ 1955 年第二届文艺复兴百货公司金圆规奖展，米兰塞尔贝洛尼宫新闻俱乐部。
1955 Exhibition of the 2nd La Rinascente Compasso d'Oro award, Circolo della Stampa, Palazzo Serbelloni, Milan.

ADI 设计博物馆档案
Archivio ADI Design Museum

▲ 1957 年第四届金圆规展，由布鲁诺·穆纳里和吉安·卡洛·奥尔泰利设计，第十一届米兰三年展。
1957 Exhibition of the 4th Compasso d'Oro award, design by Bruno Munari and Gian Carlo Ortelli, XI Triennale di Milano.

ADI 设计博物馆档案
Archivio ADI Design Museum

▲ 1960 年第六届金圆规奖展，由马里奥·贝利尼和伊塔洛·卢皮设计，米兰王宫女像柱厅。
1960 Exhibition of the 6th Compasso d'Oro award, design by Mario Bellini and Italo Lupi, Sala delle Cariatidi, Palazzo Reale, Milan.

ADI 设计博物馆档案
Archivio ADI Design Museum

▲ 1957 年文艺复兴百货公司金圆规优秀设计奖展览，由佛朗哥·阿尔比尼设计，纽约世界博览会。
1957 La Rinascente Compasso d'Oro for Good Design exhibition, design by Franco Albini, New York World's Fair.

ADI 设计博物馆档案
Archivio ADI Design Museum

▲ 1957 年文艺复兴百货公司金圆规优秀设计奖展览宣传册，由佛朗哥·阿尔比尼设计，纽约世界博览会。
1957 Brochure of the exhbition La Rinascente Compasso d'Oro for Good Design, design by Franco Albini, New York World's Fair.

ADI 设计博物馆档案
Archivio ADI Design Museum

金圆规奖的评选活动历来与时俱进，反映出社会的变迁与品位的演变。

The Compasso d'Oro events have kept reflecting the evolution of taste and society.

从诞生之日走到今天，金圆规奖探讨了不同的主题，也表彰了诸多专业人士，从最初的设计师、制造商，到如今的设计文化杰出宣传大使，不一而足。该奖项由此在设计界占据了举足轻重的地位，并借此反映出设计领域及其关注重点的变化。

From the beginning until today, different professional figures – extending from designers and manufacturers in the origin to present time with outstanding promoters of design culture – and different themes have been addressed by the award, thus channeling through a critical positioning a sense of the evolution of design world and its priorities.

▲ 1970 年第十届金圆规奖，目录。
1970 X Compasso d'Oro, catalogue.

ADI 设计博物馆档案
Archivio ADI Design Museum

▲ 1970 年第十届金圆规奖展，由伊塔洛·卢皮和罗伯托·卢奇设计，米兰斯福尔扎城堡舞厅。
1970 Exhibition of the 10th Compasso d'Oro award, design by Italo Lupi and Roberto Lucci, Sala della Balla, Castello Sforzesco, Milan.

ADI 设计博物馆档案
Archivio ADI Design Museum

▲ 1955 年，阿德里亚诺·奥利韦蒂荣获意大利金圆规奖。
1955 Adriano Olivetti receiving the Grand National Prize.

ADI 设计博物馆档案
Archivio ADI Design Museum

▲ 1958 年，阿斯格·费舍尔荣获国际金圆规奖，佛朗哥·阿尔比尼荣获意大利金圆规奖，罗马国立现代艺术美术馆。
1958 Asger Fisher receiving the Grand International Prize and Franco Albini receiving the Grand National Prize, Galleria Nazionale d'Arte Moderna, Rome.

ADI 设计博物馆档案
Archivio ADI Design Museum

▲ 1959 年，奥古斯托·莫雷洛为小埃托雷·索特萨斯颁发金圆规奖，米兰现代艺术博物馆当代艺术馆。
1959 Ettore Sottsass Jr. receiving the compasso d'Oro award from Augusto Morello, Padiglione d'Arte Contemporanea, GAM, Milan.

ADI 设计博物馆档案
Archivio ADI Design Museum

▲ 2022 年，温琴佐·卡斯泰拉纳和安德烈亚·罗瓦蒂为罗萨纳·奥兰迪颁发金圆规终身成就奖，米兰 ADI 设计博物馆。
2022 Rossana Orlandi receiving the Compasso d'Oro Carrer Award from Vincenzo Castellana and Andrea Rovatti, ADI Design Museum, Milan.

© 2022 罗伯托·德里卡尔迪斯 / 罗萨纳·奥兰迪档案
© 2022 Roberto De Riccardis / Archivio Rossana Orlandi

几十年来，多种多样的设计类别加入了金圆规奖大家庭，如研究式设计、平面设计、服务设计和装置艺术等。此外，涉及运动或饮食文化的专项奖也拓宽了金圆规奖覆盖的设计疆域。丰富的作品目录就是这一演变最直接的见证者。

Across the decades, different and multiple categories have joined the realm of the award, such as design by research, graphic design, services, and installations. In addition, special awards dedicated to sports or food culture have expanded the geography of design described by the Compasso d'Oro. A rich collection of catalogues stands as the most immediate witness of such evolutionary traits.

金圆规奖的普及范围不断扩大，从最初设计界专业人士的小众狂欢，变成了如今让广大公众参与其中，甚至同时面向整座城市的开放平台。

Compasso d'Oro has expanded from the disciplinary audience of design, to an open platform for a wider, more involved public, and for a whole city at the same time.

▲ 1979 年《设计与设计》——1954—1970 年金圆规奖展览目录，佛罗伦萨历史中心。
1979 *Design & design*, catalogue of the exhibition Compasso d'oro 1954 - 1970, Centro Di, Frienze.

ADI 设计博物馆档案
Archivio ADI Design Museum

▲ 1994 年第十七届金圆规奖，目录。
1994 XVII Compasso d'Oro, catalogue.

ADI 设计博物馆档案
Archivio ADI Design Museum

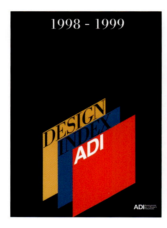

▲ 1999 年《ADI 设计索引（1998—1999 年）》，目录。
1999 Design Index ADI 1998-1999, catalogue.

ADI 设计博物馆档案
Archivio ADI Design Museum

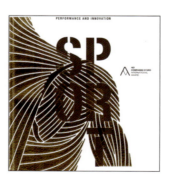

▲ 2017 年《以运动为主题的 ADI 国际金圆规奖》，目录。
2017, Sport ADI Compasso d'Oro International Award, catalogue.

ADI 设计博物馆档案
Archivio ADI Design Museum

▲ 2020 年米兰 ADI 设计博物馆开馆。
2020 ADI Design Museum opening, Milan.

摄影：吉吉·科波拉 /ADI 设计博物馆档案
Ph. Gigi Coppola / Archivio ADI Design Museum

▲ 1955 年马塞尔·布劳耶荣获国际金圆规奖的颁奖典礼，米兰塞尔贝洛尼宫新闻俱乐部。
1955 Presentation of the Grand International Prize awarded to Marcel Breuer, Circolo della Stampa, Palazzo Serbelloni, Milan.

ADI 设计博物馆档案
Archivio ADI Design Museum

金圆规奖历经 70 年的发展，积累起独一无二的历史收藏与档案，包括 2300 多件作品、丰富的出版物与历史文件，体现了该奖项巨大的文化潜力。

The seven decade-long history of Compasso d'Oro has generated an unequaled historical collection and archive, where more than 2.300 objects and a rich archive of publications and historical documents are gathered, embodying the enormous cultural potential of the award.

自 2021 年开馆以来，ADI 设计博物馆定期更换常设展览的展品，全面展出多年来积淀的丰富档案遗产，同时举办大量临展，持续着力对话类活动，与公众深入互动。

Open since 2021, ADI Design Museum is a home to the permanent collection – which is periodically refitted to fully display the rich patrimony of archival resources – and a powerful physical touchpoint with people, through significant temporary exhibitions and continuous occasions of dialogue.

▲ 2021 年"勺与城市"展览，贝佩·菲内西策展，米兰 ADI 设计博物馆。
2021 Il cucchiaio e la città (The Spoon and the City), curated by Beppe Finessi, ADI Design Museum, Milan.

ADI 设计博物馆档案
Archivio ADI Design Museum

▲ 2021 年"一对一：品类之盛"展览，贝佩·菲内西策展，米兰 ADI 设计博物馆。
2021 Uno a uno. La specie degli oggetti (One to One: the Species of Objects), curated by Beppe Finessi, ADI Design Museum, Milan.

摄影：莱奥·托里 / 莱奥·托里档案
Photograph by Leo Torri / Archivio Leo Torri

▲ 2021 年"一对一：品类之盛"展览，贝佩·菲内西策展，米兰 ADI 设计博物馆。
2021 Uno a uno. La specie degli oggetti (One to One: the Species of Objects), curated by Beppe Finessi, ADI Design Museum, Milan.

摄影：埃利斯·焦雷塔伊 /ADI 设计博物馆档案
Photograph by Elis Gjorretaj / Archivio ADI Design Museum

▲ 2022 年第二十七届金圆规奖展，米兰 ADI 设计博物馆。
2022 Exhibition of the 27th Compasso d'Oro award, ADI Design Museum, Milan.

ADI 设计博物馆档案
Archivio ADI Design Museum

▲ 2023 年博物馆中心区域的常设展和临展："当下即永恒"展览，弗兰切斯卡·巴莱纳·阿里斯塔、乔瓦尼·科莫利奥和迈特·加西亚·桑奇斯策展；"意大利：新的共同景观"展览，由安杰拉·鲁伊、伊丽莎白·多纳蒂·德孔蒂和马蒂勒·洛西策展，米兰 ADI 设计博物馆。
2023 Permanent and temporary exhibitions at the central space of the museum: Presente Permanente (Permanent Present), curated by Francesca Balena Arista, Giovanni Comoglio and Maite Garcia Sanchis; Italy A New Collective Landscape, curated by Angela Rui with Elisabetta Donati de Conti and Matile Losi, ADI Design Museum, Milan.

摄影：马丁娜·博内蒂 /ADI 设计博物馆档案
Photograph by Martina Bonetti / Archivio ADI Design Museum

# 金圆规奖的源起与发展：来自意国，世所无双

弗兰切斯卡 · 巴莱纳 · 阿里斯塔

# Compasso d'Oro:
# Origin, Evolution, and
# Uniqueness of an Italian Award

Francesca Balena Arista

金圆规奖始终与意大利设计界的重大事件和重要人物密切相关，呈现的重要设计都是一路以来已经渗入意式工业设计的肌理，持续产生影响的作品。

金圆规奖始终关注工业设计的角色作用以及工业与匠艺的相互影响。在该奖的发展史上，于1954年创立奖项的文艺复兴百货公司和1956年成立的意大利工业设计协会（ADI）是必然要提及的两个名字。另外，战后顶级设计大咖也携手开创、共同孕育了这一奖项，包括建筑师、设计师、Domus 杂志创始人吉奥·蓬蒂，Stile Industria 杂志创始人、总监阿尔贝托·罗塞利，文艺复兴百货公司艺术总监、平面设计师阿尔贝·斯坦纳。

金圆规奖自创立之日起，一直处在意式设计文化的巅峰，正如1954年第1期的 Stile Industria 杂志所说，这个奖项"开创性地认识到，生产不仅是一种经济现象，而且还涉及美学、文化与习俗问题。"

一篇名为《产品美学奖》的文章横空出世，文艺复兴百货公司借此奖项"鼓励工业企业和手工匠人不断打磨产品的技术水平和外观样式。"

最初规定，"所有意大利工业与手工艺设计公司以及模型制作者"（即设计师）均可报名参赛。战后对工业设计与设计师间的角色尚无定论，金圆规奖面向生产者和设计师两大群体，希望推动已有的讨论。

## 金圆规奖创立背景

1918年12月，文艺复兴百货公司米兰教堂广场总店开门迎客，走在当时欧洲潮流的最前端，客户中不乏那时最知名的人物。文艺复兴百货真正是领商业、文化领域一时风骚。但其目光所及，远不止产品销售，而是更着力于通过各类展览、活动塑造意式喜好。久已习惯定制家具、量身剪裁、手工制造的意大利中产阶级，在文艺复兴百货的橱窗中，渐渐熟悉了工业生产的各类产品。

此处，必须提到1953年的"产品美学"展览，即金圆规奖的前身，展出了家具、日用品、纺织品、服装、佩饰等各类商品。展览的一大亮点是背后的组织团队，卡洛·帕加尼、布鲁诺·穆纳里、阿尔贝托·罗塞利负责策展，阿尔贝·斯坦纳则贡献了优秀的平面设计。

史上首次由一家百货公司来解决产品的"广义尊严"这个"社会问题"。斯坦纳设计的展览宣传册中这样写道："无论是一把刀、一条裙子，还是一只玻璃杯、一个熨斗，产品的尊严与价值不仅体现在价格与功能上，也关乎外观形态……"

1954年，文艺复兴百货公司设立文艺复兴百货公司金圆规奖。获奖产品在文化重头戏米兰三年展上亮相。能参与如此重要的展会，获奖产品品质可见一斑。

这一年也是工业设计史上的里程碑之年。第十届米兰三年展以工业

设计为专题，科技博物馆也举办了工业设计第一届国际代表大会。同年，阿尔贝托·罗塞利创办了意大利首份工业设计专门刊物 Stile Industria 杂志，他认为工业设计是"生产与文化和谐关系"的结晶。这份杂志在工业设计整体面貌和流行品味的塑造中扮演了关键角色。

## 早期金圆规奖与 ADI 的诞生

直至1957年，前四届金圆规奖均由文艺复兴百货公司举办、管理。1956年，在吉奥·蓬蒂、阿尔贝托·罗塞利等知名设计师的推动和努力下，意大利工业设计协会（ADI）正式成立，罗塞利任首届主席。ADI 自1958/59年开始参与金圆规奖的筹办运营，在合作数年之后，自1967年起，完全接管奖项的管理工作。

自金圆规奖创立至1960年，各分奖项略有差异：制造商获颁金圆规奖；设计师获颁银质荣誉奖牌和现金奖励；荣誉提名奖则颁发证书。自1964年起，向机构和个人获奖者统一颁发金圆规奖，进一步巩固了金圆规的重大象征意义。金圆规奖的标志由阿尔贝·斯坦纳设计，经由阿尔贝托·罗塞利和马尔科·扎努索设计转化为具体实物。

1955年的第二届金圆规奖包括两大奖项——意大利金圆规奖授予阿德里亚诺·奥利韦蒂；国际金圆规奖则颁发给马塞尔·布劳耶。奖项不仅是对优秀产品的认可，也是对为设计领域发展做出卓越贡献的个人、机构、组织的嘉奖。奖项门类的扩充，表明对推动设计发展的各类因素予以认可与肯定。从金圆规奖设立伊始，这样的转变就显而易见。比如，1954年首届颁奖典礼上，评委会在对获奖作品——Olivetti "字母22"打字机的评价说明里特别指出，"这一奖项是对奥利韦蒂公司杰出工作的认可。奥利韦蒂公司是各大设计门类相互协作、交相辉映的全球领军代表。"

第三届金圆规奖评选中，意大利金圆规奖授予吉奥·蓬蒂，国际金圆规奖则花落纽约现代艺术博物馆。后来数年间，获奖者包括重要的设计师个人和机构，着力激励设计领域内的交流合作关系。渐渐地，金圆规奖走向了国际舞台。例如，1957年的纽约世界博览会推出了"文艺复兴百货公司金圆规优秀设计奖"专展。公众得以近距离欣赏获奖产品，领略意式设计文化的精髓。

1994年，意大利/国际金圆规奖融合成为金圆规终身成就奖。2020年，增设了金圆规产品终身成就奖。

历届金圆规奖都会同时举办作品展览，均由知名设计师操刀主持。金圆规奖始终致力于提高公众对设计的认识，在优秀机构和重要文化场所呈现一场又一场令人难忘的展览。比如，第六届金圆规奖展就精彩纷呈，由马里奥·贝利尼和伊塔洛·卢皮策展，在米兰王宫女像柱厅举办。

## 金圆规奖：步履不停

1967 年，ADI 开始完全接管奖项的管理工作，与金圆规奖一起见证品味和社会的变迁，而这一奖项也有自己坚定的姿态。由此，金圆规奖就是一部设计学科的发展史，一路上铸造了一座又一座里程碑，也不断应对一个又一个结构性挑战。

1958 年以前，金圆规奖参赛作品只能由各家百货公司推荐。在 ADI 的努力下，参赛资格范围最终扩大到所有生产部门。此后的金圆规奖逐步发展成为推动文化发展的重要力量，涵盖推进工业设计进步的理论研究和设计项目。

每届金圆规奖都会结集出版作品集，集中展示获奖作品，也激发观众的深入思考。在 1970 年第十届获奖作品集中，ADI 时任主席安娜·卡斯泰利·费里埃里提出要吸引更多的观众。这届金圆规奖恰在首届米兰设计周期间举办，出台了奖项新规，从实践和挑战两个维度更加全面地展现意式设计的风貌。卡斯泰利·费里埃里主张在设计中理论结合实践，强调思辨性作品的重要意义。同时，她指出"生产体系面临的挑战，已经在消费主义的漩涡中越陷越深。"尤其引人注意的是，本届评委会中有法国社会学家、《物体系》作者让·鲍德里亚。另外，有趣的是，埃托雷·索特萨斯设计的 Olivetti "MC 19"电子加法计算器荣获殊荣；而知名度更高的"情人"打字机，虽同样出自索特萨斯之手，却只获得了荣誉提名奖。

研读评委给出的获奖说明，可以一窥历史大背景下的设计领域，了解金圆规奖的评奖标准。1981 年的金圆规奖《总评审报告》提出"研究"分项，强调"需要增强科技与生产相关的研究，不断换新表达语言。"评委会挑选出 Driade、Alchimia、Zanussi 三家公司，认为他们能够彰显意大利工业的品质与创造性。Alchimia 的 Controdesign 质疑传统的设计师—设计产业关系，而 Driade 和 Zanussi 的设计恰恰以此种传统关系为依托；尽管三家公司的模式不同，但并不妨碍他们各自精彩。Alchimia 对表达语言的更新换代，作用尤大，对生产方式也具有重大影响，开拓了自我生产的新路径。

2022 年，我们举办了 ADI 设计博物馆第二届"当下即永恒"展览，联手迈特·加西亚·桑奇斯和乔万尼·科莫利奥，设计三大主题区：有效解决所处时代需求的设计、预见未来需求的先锋设计、挑战所处时代流行范式的"宣言式"设计，充分展示设计与所处时代背景间错综复杂的纠葛。每件展品都紧密镶嵌在周围环境当中，直面紧迫的时代之间。在筹备本次新展中，我们依然在思考，金圆规奖历经七十年，对展现设计界多彩多样面貌的贡献。这种杰出贡献有赖于历届表现卓越的评审委员会——他们始终坚持评审结构的多样性、多元化，涉及各类主题，设定多个评审级别。已经举办了 27 届的金圆规奖评选，依然不改初衷，努力呈现设计的多元面貌，涵盖当代社会重要的文化、经济维度。

The Compasso d'Oro Award is intricately linked to significant events and figures of Italian design: representing the primary subjects of discourse that have permeated and continue to influence the trajectory of industrial design within the Italian framework.

In this narrative, significant contributors include La Rinascente, an innovative department store that instituted the award in 1954, and ADI, the Italian Association for Industrial Design, which was established in 1956. The discourse revolves around the role of industrial design and the interplay between industry and craftsmanship. The individuals involved in conceiving and nurturing the award are widely acknowledged as among the most influential figures in the post-war era. They include the architect and designer Gio Ponti, the founding force behind *Domus* magazine, Alberto Rosselli, the originator and director of *Stile Industria* magazine, and the graphic designer Albe Steiner, who served as the art director of La Rinascente.

The award has consistently represented the pinnacle of Italian design culture since its inception. It can be defined as "the first event that recognised and framed the issue of production as not only an economic but also an aesthetic, cultural, and behavioral matter," as stated in the presentation that appeared in the *Stile Industria* magazine's first issue in 1954.

An article with the headline *Aesthetics in the Product* was published, which was established by La Rinascente "to encourage industrialists and artisans to enhance their production not only technically but also in terms of form."

The initial regulations of the award stipulate that "all Italian industrial and artisanal companies and the creators of models" (i.e., designers) can participate. The role of industrial design and the designer was still in the process of definition in the post-war era. The birth of the award, presented to both producers and designers, aimed to encourage the debate.

## Background and Establishment of the Award

La Rinascente unveiled its Piazza Duomo headquarters in Milan in December 1918. The store sought inspiration from the most cutting-edge European trends and engaged with some of the most prominent figures of that era, evolving into a genuine focal point, not solely in commerce but also in cultural circles. Its purpose extended beyond product sales; it aimed to shape the preferences of Italians, achieved through exhibitions and events. This approach gradually introduced the Italian middle class, traditionally accustomed to bespoke furniture, tailored attire, and artisanal creations, to mass-produced items displayed in the store's showcases.

In this historical context, it is crucial to acknowledge the 1953 exhibition, *Aesthetics in the Product*, as a precursor to the award. This exhibition included a curated collection of items available for purchase, encompassing furniture, household articles, textiles, apparel, and accessories. Notably, the exhibition's design was orchestrated by Carlo Pagani, Bruno Munari, and Alberto Rosselli, with graphic design masterfully handled by Albe Steiner.

For the first time, a department store addresses the "social issue" of the "greater dignity" of the product. "A knife, a dress, a glass, an iron, have their dignity not only in the price or function, but also in the design (...)" reads the presentation leaflet, designed by Steiner.

In the following year, 1954, La Rinascente established the *Compasso d'Oro La Rinascente Award*. The winning products were showcased at the Milan Triennale, a significant institution actively engaged in cultural discourse. The quality of the selected products was emphasised through the quality of the exhibition.

The year 1954 marked a pivotal moment in the discourse on industrial design. The 10th edition of the Milan Triennale was exclusively dedicated to this theme, and the Museum of Science and Technology hosted the First International Congress on the subject. During this crucial period, Alberto Rosselli founded *Stile Industria*, the very first magazine in Italy entirely devoted to this domain. In Rosselli's perspective, industrial design is the outcome of a "harmonious relationship between production and culture." His magazine played a pivotal role in molding the discourse and influencing prevailing tastes.

## The Early Editions and the Birth of ADI

La Rinascente handled the production and management of the initial four editions of the award until 1957. The Italian Association for Industrial Design (ADI) was established in 1956, with notable figures such as Gio Ponti and Alberto Rosselli among its promoters and founders. Alberto Rosselli assumed the role of the first president of ADI. ADI became involved in the award in 1958/59, and after a collaborative partnership spanning several years, assumed full management in 1967.

Since the award's establishment until 1960, a differentiation existed in the recognitions: manufacturers were honored with the Compasso d'Oro, while designers received a Silver Honorary Plaque along with a cash prize. Diplomas were presented for Honorable Mentions. Starting from 1964, both companies and designers were recipients of the Compasso d'Oro, solidifying the symbolic significance of this item, which was conceived by Albe Steiner and transformed into a three-dimensional object by Alberto Rosselli and Marco Zanuso.

In 1955, the second edition of the award introduced two significant categories: The National Grand Prize, which was awarded to Adriano Olivetti, and the International Grand Prize,

presented to Marcel Breuer. These Grand Prizes aimed to recognise not only individual products but also the outstanding contributions of individuals, institutions, or organisations to the field of design. This expansion in award categories signified the need to acknowledge and celebrate the broader impact of design endeavors. This shift was evident right from the award's inception. For example, in the very first edition, when the jury honored the Olivetti Lettera 22 typewriter, they explicitly emphasised that "It can be symbolically considered as extended to the exceptional merits of Olivetti's entire activity, which represents globally one of the highest examples of stylistic coherence of an entire production in all the accompanying expressions..."

In the third edition, the National Grand Prize was awarded to Gio Ponti and the international one to the MoMA in New York. In the following years, awards were given to significant individuals and institutions, emphasising the creation of relationships for exchange and collaboration. The opening to the international context was underscored by events such as the 1957 New York World's Fair, featuring the exhibition *La Rinascente Compasso d'Oro for Good Design*. The public was directly engaged, able to closely examine the presented products, representing a real introduction to Italian design culture.

From 1994, the Grand Prize was identified as the Compasso d'Oro Career Achievement Award. In 2020, the Compasso d'Oro Career Achievement Award for the Product was introduced.

The award editions have consistently featured accompanying exhibitions, with prominent designers overseeing the installations. Since its inception, the award's significance has rested on its role as a vehicle for reaching an increasingly broad audience. Distinguished institutions and significant cultural venues serve as the backdrop for the showcased projects, which are presented in memorable exhibitions. For instance, the sixth edition saw a remarkable exhibition conceived by Mario Bellini and Italo Lupi, hosted in the Sala delle Cariatidi at Palazzo Reale in Milan in 1960.

## An Ever-Evolving Award

In 1967, ADI took over full management of the award. The Compasso d'Oro has been a companion, reflecting the shifts in taste and society, while also taking critical stances. Consequently, its history intertwines with the evolution of the design discipline, the milestones it has achieved, and the structural challenges it has faced.

Until 1958, the award was exclusively for products distributed through department stores. However, with ADI's involvement, the eligibility criteria broadened to encompass all sectors of production. With each subsequent edition, it evolved into a more comprehensive tool for cultural promotion, encompassing theoretical studies and initiatives aimed at organising and promoting industrial design.

Each edition is accompanied by a publication, which serves as a culminating aspect of the award and has the potential to inspire contemplation. In the catalog of the 1970 edition - the tenth - the then-president of ADI, Anna Castelli Ferrieri, introduced a new direction aimed at engaging a broader audience. This edition was part of the events of Milan's inaugural Design Week. With updated regulations, it presented a more comprehensive overview of Italian design, encompassing both practical and challenging aspects. Castelli Ferrieri advocated for the integration of theoretical endeavors with practical implementation, emphasising the significance of critical work. She also concluded by addressing the "challenges to the production system, which today has become deeply entangled in a consumeristic spiral". Notably, the judging panel included French sociologist Jean Baudrillard, the author of *The System of Objects*. In an interesting aside, the jury awarded the Olivetti *MC 19* electric adding machine designed by Ettore Sottsass, while the more famous Valentine, also designed by Sottsass, received only an Honorable Mention.

The reading of the justifications for which the Juries assign the awards is an interesting tool of analysis that provides us with an interpretative key and places the work of the jurors in historical context. In 1981, in the General Jury Report, we read that in this edition the category "research" is introduced, which refers to the "need to stimulate further development of research experiences related both to the evolution of technological-production aspects and to the renewal of expressive languages." The jury selects three companies, Driade, Alchimia, and Zanussi, expressions of equally different models, capable of highlighting the quality and creativity of Italian industry. The experience of *Alchimia's Controdesign*, which questions the traditional relationship between designers and industry, is awarded as much as the work of Driade and Zanussi, which are based on this relationship. Alchimia's contribution to the renewal of expressive languages is significant, as well as its role in renewing production aspects, paving the way for self-production.

In 2022, while curating the exhibition *Presente Permanente*, the 2nd Edition of the ADI Design Museum, together with Maite García Sanchis and Giovanni Comoglio, we identified three distinct categories within the collection: projects that effectively responded to the demands of their time; pioneering projects that foresaw these demands; and manifesto projects, often challenging the prevailing norms of their era. All of these categories bear witness to the intricate relationship between design and its contemporary context. Each project reflects a complex engagement with the surrounding context and the pressing questions of the time. In presenting this new exhibition on the Compasso d'Oro Award, we once again reflect on how, over its seventy-year history, the award has played a significant role in navigating the complexities of the design world. It has accomplished this through its distinguished juries, consistently featuring significant figures from both the national and international design landscape, and through a complex system involving various thematic committees and multiple levels of evaluation. Now, as it celebrates its XXVII edition, the Compasso d'Oro Award continues to stand as an essential tool for interpreting the multifaceted phenomena of design, encompassing both cultural and economic dimensions, which remain indispensable in our contemporary society

# 嘉奖设计：展望

乔万尼·科莫利奥

# Awarding Design:
# A Visionary Approach

Giovanni Comoglio

煮咖啡过程中用到的钢制构件——发现构件的与众不同，把构件摆上展示台，这种做法到底与收藏文化的历史发展有无关联？

收藏，从来都具有多面性。四处探险，金屋藏宝者有之；求索科学，采撷自然者有之；为彰显自身文化社会地位者有之；纯粹出于个人喜好赏玩者亦有之。形形色色的收藏，无外乎收藏者的眼光或是认知所及，折射出的通常也不过是当时的历史背景，远非历史本身。

金圆规奖历史性收藏包括自奖项 1954 年设立起至 2023 年间的 379 件获奖作品和 1192 件提名奖作品，以奖项设立七十年来的独特风格，傲视群雄；于 2004 年 4 月被意大利文化部正式纳入意大利文化遗产范围。与其他收藏不同，本收藏不是对某段历史时间或主题性碎片的封藏，亦不受历史事件或策展标准的约束，而是仿若一条以 1954 年首届藏品为起点的射线，走向路径没有定式，只有在作品集中凝结下来的历史痕迹。金圆规奖是当下发展进程的固有部分，攫住眼前的一个又一个瞬间。

这种独特气质源自金圆规奖与意式设计诸多门类间的复杂关联。奖项的初衷本就是界定并凸显战后意式设计的重要意义，由最初的产品设计逐步扩展到设计在当代社会各个领域中的内在价值。金圆规奖由文艺复兴百货公司市场调研办公室发起，植根于意大利工业、文化的沃土，得到众多重量级人物的加持，如 *Stile Industria* 杂志的阿尔贝托·罗塞利，以及在 *Domus* 杂志耕耘近 30 年、始终致力于推行"由小家见文化"观点的吉奥·蓬蒂。1953 年，文艺复兴百货公司举办"产品美学"展览，即金圆规奖前身，次年进行首次评奖，希望推动全社会对设计的反思，强调要把物质文化置于对当前和历史的建构之中。

1956 年，意大利工业设计协会成立，后于 1958 年开始为金圆规奖提供支持，融合设计领域涉及的文化、构思、生产、发行等方面，最终正式形成现在的设计体系。

金圆规奖从丰富的源头中汲取能量，独具特色。不同于博物馆藏品对过去的总结与回顾，金圆规奖着眼于当下，既捕捉现今的倾向与趋势，也留住在历史长河中被时间冲刷掉的主题和视角。

当下和历史都在金圆规奖作品中永固，经由历届国际评委精挑细选而来。同时，恰恰因为是所处时代的一分子，评选人也为时代留下了特有的"透视照"。就像 2008 年的金圆规奖国际评委会为了凸显奖项的择优性，特意将获奖名额从 20 个减缩至 10 个。奖项评选充分显示出金圆规奖及其评委会不僵化、不泥古的特质。

亚历山德罗·门迪尼担任艺术指导的 Alessi 推出了茶具／咖啡具系列"茶与咖啡露天广场"，阿尔多·罗西认为这个系列的设计是建筑艺术与餐具的完美融合，向着即将到来的后现代时期高歌猛进。但系列中仅有理查德·迈耶的设计获得了 1984 年的金圆规奖，评委会认为迈耶的设计"植根于丰富多样的现代运动传统，丝毫没有泥古的时代倒退。"1962 年，卡斯蒂廖尼设计的灯具"拱"和"皮塔戈拉"咖啡机同时参选，但只有咖啡机获得了当年的金圆规奖。这样的评选

结果并非谬误——"皮塔戈拉"是时代的结晶，体现着当时的价值观念。"拱"看似湮灭在共同想象的历史里，然而却在来日方长中慢慢成为偶像式的存在。

随着时间推移，奖项设置不断发展变化。例如，过去的本土大奖和国际大奖合并成为终身成就奖，截至目前共颁发 152 项；同时，自 2020 年开始，也向长盛不衰的产品颁发终身成就奖。藉此，金圆规奖实现了对过往优秀设计师和设计作品的追溯回顾。尽管如此，获奖的追溯性设计作品或多或少与现在存在关联，能够映射出设计作品符号化过程中的特点，体现符号化设计产生的历程和其中蕴含的多方努力。

金圆规奖始终以多种方式体现、描摹当下，各类相关领域和研究活动交叉汇集。金圆规奖藏品的性质、建构模式和发展壮大，无一不闪烁着设计与科技、人文交相辉映的光芒，佐证了设计和其他学科共有的纷繁复杂，集聚了设计、生产、销售、消费等各方努力，具有不可估量的文化社会影响，也擘画出设计师、建筑师、工业人士、评论界、技师、商人、广告界、平面设计师、商品展陈、新闻记者等众多设计体系相关从业者的路径。雷纳托·德·富斯科在 ADI 历史的专著中对此有专门论述。

这样的交叉汇集首先体现在对获奖设计作品相关档案、背景资料的收集，始自 ADI 于 1958—1959 年间开始参与金圆规奖，最终形成了现在的 ADI 历史档案。乔瓦尼·萨基在意式设计与国际设计领域深耕数十载，ADI 历史档案取址乔瓦尼·萨基工作室，充分体现出档案与设计史滚滚车轮的牵绊。

20 世纪末，对金圆规奖和 ADI 在形成设计体系框架中的角色作用有了更加深刻的理解。意大利常任设计观察组成立，监督管理意式设计年度出版物《ADI 设计索引》的编辑出版工作，既能更好展现当今设计领域的复杂图景，又保有了金圆规奖的独特魅力。近年来，参赛作品数量显著增加；从 5000 份左右参赛作品中挑选出大约 700 件，汇编成两版《ADI 设计索引》，而最终获得金圆规奖殊荣的作品不超过 20 件。

当然，不断发展壮大的 ADI 藏品最重要的闪光点还是"人"。六十多年来，围绕藏品地点的选取，讨论不断，类似"博物馆"的语汇经常出现，但更多的是"接触点"这个词。藏品展览始终是为了增强人与获奖作品间的联结。

这个地点从来都不是一成不变的。自上世纪 50 年代起，与藏品相关的展览始终强调人与获奖作品实质性之间的联结；或许，从某个时刻起，获奖作品本身已经没有那么重要——从 1954 年的米兰三年展到 1957 年的纽约世界博览会，再到米兰家具展和厨房、椅子，共展出了 150 件藏品。

展览历年来选取博物馆、学校、机构等地举办，在互联网时代，甚至改为网上办展，直到 2010 年，落户米兰，ADI 设计博物馆最终在

2021 年正式开馆。

ADI 设计博物馆同样以藏品的历史社会作用为主线，设计展览空间。博物馆是米兰的一个城市联络点，向公众开放，是米兰不同领域的交汇平台，博物馆的展览布局更是一份宣言：充分利用每一寸可用空间，展示 ADI 藏品，也举办临时展览、单年 / 双年"设计索引"和金圆规展览、会议、工坊，让公众在连续而流动的公共空间中徜徉。

本次展览展示了历届获奖者的作品，并从每届中挑选一个设计深入阐释。展览对金圆规奖的获奖作品"断代"再选，重新审视，经由不断发展的藏品主题，与个体和整个设计体系展开思辨性对话。我与弗兰切斯卡·巴莱纳·阿里斯塔、迈特·加西亚·桑奇斯共同策展的"当下即永恒"展览希望传达这样一种观念：通过作品描述、评审合规、现场讨论，凸显展览选取的历届获奖作品在所属时代的角色作用，以及在当代的持续性影响。

一言蔽之，此次展览凝结了金圆规奖的菁华，展现金圆规奖藏品的繁杂多样、独一无二、无可替代。

Does recognising and showcasing specific steel components used in the coffee-making process hold any connection to the historical development of a collective culture?

Throughout history, collections have played diverse roles, from Wunderkammer, a fusion of wonder and exploration, to scientific collections demonstrating the natural world, to collections aimed at reinforcing the cultural or social status of individuals and institutions, and even those created solely for the enjoyment of individuals. These collections are typically the result of a particular vision or interpretation, often reflecting the historical context, rather than history itself.

The Compasso d'Oro historical collection, encompassing 379 awardees and 1192 mentions as of 2023, stands out due to its uniqueness and its continuous evolution. It is a collection in continuous evolution commenced in 1954 with the award's first edition and was officially recognized by the Ministry of Culture in April 2004. What makes it even more distinctive is that its progression remains indefinite, with only the 1954 inception serving as a fixed point. Using geometric nomenclature as an aid, while other collections often encapsulate temporal or thematic segments defined by their story or by curatorial choices, the Compasso d'Oro collection can be likened to a semi-straight line. Its inception in 1954 serves as the solitary fixed point, and its progression remains indefinable except in the precise moment when its points appear on paper. This is because the prize itself is, and continues to be, by its very nature, part of the generative process of the present: it identifies a sequence of snapshots of the present.

This unique nature is intertwined with its historical connection to various trajectories within Italian design. Its origin speaks of a post-war Italy which felt the need to define and acknowledge

design—initially product design—as a distinct field with inherent value in contemporary society. The endeavour was initiated by the La Rinascente department store's research office, but it was rooted in a broader trend within the Italian industrial and cultural context. Figures like Alberto Rosselli, with the magazine *Stile Industria*, and especially Gio Ponti, who had already been promoting for almost thirty years, with his *Domus*, an idea of home as a human habitat going far beyond its building structures to become a mirror of culture in the broadest sense, especially through the material landscape of objects that contributed to its constitution. In 1953, the exhibition *L'estetica nel prodotto* (Aesthetics in the Product), promoted by Rinascente and preceding the award, as well as the following year's first award edition, aimed to share this reflection with society. They emphasised the need for a connection between material culture and the construction of the present and history.

In 1958, ADI (the Industrial Design Association), founded in 1956, joined the process, offering its support and formalising the need for systematisation by uniting cultural, ideational, production, and distribution forces that collectively form what we now recognise as the design system.

This layered baggage of origins leads to a unique outcome: unlike collections constructed retrospectively or as museum or curatorial endeavours, the Compasso d'Oro crystallises the present moment and, in doing so, it captures not only its partialities, but also themes and perspectives that the chronological passage of history would naturally remove.

This phenomenon is mirrored by the award itself, its juries, their choices and motivations by being part of the social, cultural and economic context of their time. They have helped to fix an extremely specific x-ray of it. In 2008, for instance, the international jury opted to limit the awards to just 10 out of a possible 20, in a deliberate act of signalling through their selectivity. Other instances of positioning similarly illustrate this dynamic.

In 1984, *Tea and Coffee Piazza*, a series of tea and coffee sets by the Alessandro-Mendini-art-directed Alessi was awarded, erasing in an idea by Aldo Rossi all distances between architecture and tableware, playing a triumphal march for the incoming postmodern era; the jury, however, awarded only one set, the one designed by Richard Meier, which at that time showed having "its roots in the vast, multidimensional heritage of the modern movement, avoiding retrograde anachronistic historical citations". In 1962, a choice was made between two products endorsed by Castiglioni: the *Arco* lamp and the Pitagora coffee machine. Interestingly, the *Pitagora* coffee machine was selected to receive the Compasso d'Oro award, which can be seen not as an error but rather as a reflection of the prevailing values and context of that time. Subsequently, the *Arco* lamp became iconic in the public consciousness, but its journey to iconic status unfolded in the years that followed.

With some time having transpired, the reconfiguration of individual accolades, including the transformation of past national and international grand prizes into lifetime achievement awards (totalling 152 to date), and the recent addition of long-selling product lifetime achievement awards (commencing in 2020), provides opportunities for retrospection. Yet, it portrays something that is a relation to the present, to an enduring dimension particularly concerning the processes of iconisation and the creation of icons, which are inherently collective endeavours.

More broadly than the prize itself, the ways of embodying and portraying the present have proven over time to be manifold, encompassing a wide geography of relational and research interfaces. In its nature, mode of construction, and growth, the Compasso d'Oro collection indeed embraces an enormous complexity. This complexity is a fundamental and distinctive characteristic of design and other disciplines, found at the intersection of technology and humanism, where human life unfolds and develops. It exists at the junction of various contributions, including the stages of design-production-sales-consumption, but also countless cultural and social implications, and diverse pathways such as those of designers, architects, industrialists, critics, factory technicians, marketers, advertisers, graphic designers, showroom managers, journalists, and many others, as evoked by Renato De Fusco in his history of ADI.

First of these interfaces is the initiation of an additional collection effort that accompanied the collection of the award-winning projects, involving gathering documents related to the awarded designs, particularly focusing on their context. This initiative has been ongoing since the involvement of ADI with the Compasso d'Oro in 1958-1959, and it led to the creation of the present-day ADI historical archive. Furthermore, the archive location, where Giovanni Sacchi, a prominent model maker for Italian and international design, conducted his work for decades, underscores its deep involvement in the flow of such ever-evolving history.

By the end of the millennium, a more formalised understanding of the role played by the combination of the Compasso d'Oro and ADI in shaping the framework of a system had already been achieved. The establishment of the Permanent Observatory on Design marked a significant development, overseeing the ADI Design Index - an annual compilation showcasing selected Italian design. This expanded the capacity to depict an increasingly complex present while maintaining the prize's exclusivity. In recent years, participation has grown substantially, with around 5,000 candidates considered, of which approximately 700 are chosen for inclusion in two editions of the Index. Ultimately, only a maximum of 20 designs are honoured with the prestigious Compasso d'Oro award.

Yet, the most significant aspect of this ever-evolving collection is where it engages with people. For over six decades, discussions have revolved around a physical place to house the collection, often involving terms like "museum," but it's more aptly described as a "touchpoint". Exhibitions related to the collection have played a vital role in reaffirming the connection between people and the material presence of award-winning projects.

It may not be a unified place, as it has long been the case. The collection-related exhibitions, from the 1950s on, have taken on this role, of reaffirming the contact between people and materiality – or less material presence, from a certain point onward – of the award-winning projects, from the 1954 exhibition at Milan Triennale to 1957 New York World's Fair, to the shows that Salone del Mobile devoted in Milan to the kitchen or chair, where the collection was present with 150 pieces.

However, the discourse has predominantly revolved around a central location, assuming various designations over time – museum, school, centre – even entertaining the notion of complete online dematerialisation during the internet boom years, only to revert in 2010 to a physical site, agreed upon with the city of Milan. This culminated, following a design competition and further refinement, in the opening of the ADI Design Museum in 2021.

Once again, it is the role of the collection in history and society that shapes the character of the spaces, rendering this museum a touchpoint. Beyond serving as an urban connector, an integral part of the city's fabric accessible as public space, open onto a square connecting different areas of a changing Milan, the exhibition layout itself serves as a manifesto: by rearranging the collection across the entire available space and integrating it with temporary exhibitions, the annual and biennial Index and Compasso Exhibitions, as well as meetings and workshops, it creates a continuous and fluid environment, anchoring us in the present, with which we continue to evolve.

The displayed collection features all winners from each edition and delves deeper into one project from each. Regular revisiting and re-selection of these insights offer opportunities for critical dialogue with individuals and the design system, utilising the themes generated by the growing collection. Reflecting on the curation work undertaken by Francesca Balena Arista, Maite Garcia Sanchis, and myself, in the *Presente Permanente* (Permanent Present) exhibition, we aimed to explore this concept: the selected awards over the years have highlighted the role objects played in their present – in description, compliance, and open critique – and how they continue to exert such role of influence in the present as active forces.

In essence, it encapsulates the very essence of the Compasso d'Oro and what makes its collection a complex and uniquely irreplaceable entity.

# 时间线：
# 项目与趋势
# 主题展区

# Timeline:
# A Thematic Landscape
# of Projects and Trends

本展区精选了 100 多件展品以及项目，带领观众畅游意式设计 70 年的历史长河，结识 350 多位获奖者，领略历届获奖作品的风采，感受金圆规奖发展史上的高光时刻。

观众将在时间线沿途经过七大主题分展区，对凸显意式设计本质与演变的具体主题展开深入的研究和探讨，追寻意式设计的本真所在，感受意式设计与时俱进的步伐。每一个分展区均阐述了相关的社会背景，由设计师和相关从业者娓娓讲述项目与作品背后的故事。

The journey through seventy years of Italian design, encompassing over 350 Compasso d'Oro award winners, is presented here with a selection of more than 100 objects and projects, representing specific moments on the award's timeline, corresponding to the different editions.

Along the way, seven thematic insights are also punctuating the timeline flow, expanding its investigation by diving deep into specific topics characterizing the nature and evolution of Italian design, in its intrinsic identity and in its constant exchange with an ever-changing world. Each insight is a story told through several stories, of projects, designers, practitioners and social contexts.

⋀ 1954 年荣誉提名奖，
银餐具三件套
吉奥 · 蓬蒂，Arthur Krupp
银餐具套组

装点餐桌的别样风景，式样新潮，与时俱进。

⋀ Honorable Mention 1954,
Three piece silver cutlery set
Gio Ponti, Arthur Krupp
Silver cutlery set

An innovative view on housewares and decoration, receptive of the formal progress of its time.

⋀ 1957 年，米雷拉
马尔切洛 · 尼佐利，Necchi，1953 年
缝纫机

这台缝纫机不走寻常路，一改传统家用"机器"的庞大笨重，凸显了工业与设计师携手合作的重要性。

⋀ 1957, Mirella
Marcello Nizzoli, Necchi, 1953
Sewing machine

The rejection of the formal canons used in domestic 'machines' reveals the significance of this high level example of collaboration between industry and designer.

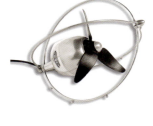

⋀ 1954 年，Zerowatt V.E. 505,
埃齐奥 · 皮拉利，Fabbriche Elettrotecniche Riunite，1953 年
风扇

直接脱胎于技术而兼具设计美学的典范。

⋀ 1954, Zerowatt V.E. 505,
Ezio Pirali, Fabbriche Elettrotecniche Riunite, 1953
Fan

A typical example of an aesthetic result deriving directly from a technical context.

⋀ 1955 年，路易莎
佛朗哥 · 阿尔比尼，Figli di Carlo Poggi，Cassina 公司自 2008 年起生产，Cassina 公司 iMaestri 系列藏品
椅

通过拼接材料，形成简洁凝练的设计方案，并保持了一贯的形式风格。设计师多年如一日地倾注心血，精益求精，不断打磨最完美的作品。

⋀ 1955, Luisa
Franco Albini, Figli di Carlo Poggi, production Cassina since 2008, Cassina iMaestri Collection
Chair

The elementary solutions (and) the inherent formality of the grafts of the material, (...) the constancy and commitment that the designer has demonstrated for years in constantly modifying and revising his work.

⋀ 1954 年荣誉提名奖，548，Balma & Capoduri 技术办公室，
Balma & Capoduri，1953 年
钳式订书机

趁手好用，功能性突出，订书机中的"战斗机"。

⋀ Honorable Mention 1954, 548, Balma & Capoduri Technical Office, Balma & Capoduri, 1953
Pincer stapler

An intuitive and functional device, becoming a reference for its category.

⋀ 1956 年，数字 5
吉诺 · 瓦莱和纳尼 · 瓦莱、约翰 · 迈尔和米凯莱 · 普罗温恰利，Solari，1954 年
电子机械时钟

灵活适应多种环境，精心选择美观的字体，专注于借助最单一的形式传递最易读的信息。

⋀ 1956, Cifra 5
Gino and Nani Valle, John Myer and Michele Provinciali, Solari, 1954
Electro-mechanical clock

Focused on achieving the greatest readability in a single format (with) adaptability (...) and attention paid to the lettering.

⋀ 1964 年，Sferyclock
鲁道夫·博内托，Fratelli Borletti，1963 年
闹钟

低成本、大销量消费品生产流通极限范围内，在设计领域精益求精的巅峰之作。

⋀ 1964, Sferyclock
Rodolfo Bonetto, Fratelli Borletti, 1963
Alarm clock

It is the result of a rigorous process in the field of re-design, (...) within the precise limits of production and distribution of a low-cost and extremely widespread consumer item.

⋀ 1964 年，紧凑型
马西莫·维涅里，A.R.P.E. Articoli Plastici Elettrici，1964 年
三聚氰胺餐具
亚历山德罗·佩德雷蒂设计收藏，米兰

叠加与紧凑的新组合，兼顾实用功能与设计美感，既有深度又有多样性。

⋀ 1964, Compact
Massimo Vignelli, A.R.P.E. Articoli Plastici Elettrici, 1964
Melamine table service
Alessandro Pedretti design collection, Milano

A new look at superimposition and concentration. (...) The communicativeness of the functional characteristics is connected to a nobility of design going further in depth and diversification.

⋀ 1960 年，T.12 Palini
阿希尔·卡斯蒂廖尼和皮耶尔·贾科莫·卡斯蒂廖尼、路易吉·卡西亚·多米尼奥尼，Palini，1960 年
课椅

旨在供意大利教育系统更新设备时为大量学生设计的一件作品。大小合适、可以叠放，简单朴素的设计恰到好处。

⋀ 1960, T.12 Palini
Achille and Pier Giacomo Castiglioni, Luigi Caccia Dominioni, Palini, 1960
School chair

An object intended for usage in large numbers as part of the renewal of the Italian education system. Perfectly sized, stackable and with the correct technological simplicity.

⋀ 1959 年，500
但丁·贾科萨，菲亚特，1957 年
汽车

车身外观一方面采用了工业与机械大批量生产的常见技术，另一方面迎合了大众消费喜好所必须考虑的经济需求，是汽车界的典范。在通过技术传达品牌本真的道路上，这款车的设计理念标志着向前迈出的重要一步。

⋀ 1959, 500
Dante Giacosa, Fiat, 1957
Car

A typical example in the automotive field of a shape resulting from the close integration of techniques typical of large scale industrial and mechanical mass production and the particular economic needs inherent in the production of a vehicle intended for mass popular consumption (...) this concepts (...) marks an important step forward on the road towards a new expressive level of authenticity within this technique.

⋀ 1967 年，蜘蛛
乔·科隆博，Oluce，1967 年
灯
亚历山德罗·佩德雷蒂设计收藏，米兰

简单调整支架角度，就能完美满足桌面、墙壁、天花板的各类照明需求。

⋀ 1967, Spider
Joe Colombo, Oluce, 1965
Lamp
Alessandro Pedretti design collection, Milano

It solves the problem of illuminating a table, wall or ceiling with the exact same lighting element, through an easy adjustment of the support mechanism.

⋀ 1970 年，MC 19
小埃托雷 · 索特萨斯和汉斯 · 冯 · 克里尔，
Olivetti，1970 年
电子加法计算器
亚历山德罗 · 佩德雷蒂设计收藏，米兰

功能性类型产品的极致方案，颜值惊人。

⋀ 1970, MC 19
Ettore Sottsass jr. with Hans von Klier,
Olivetti, 1970
Electric adding machine
Alessandro Pedretti design collection,
Milano

An extraordinary solution that is not only
functional and typological, but with an
incredibly high image quality.

⋀ 1970 年，索里亚娜
阿弗拉 · 斯卡帕和托比亚 · 斯卡帕，
Cassina，1970 年
软垫家具系列

造型简洁、线条流畅，兼具结构设计和技术
水准，简洁但不简单。

⋀ 1970, Soriana
Afra and Tobia Scarpa, Cassina, 1970
Series of upholstered furniture

For the complexity of the image achieved
with constructive and technical means of
considerable simplicity and coherence.

⋀ 1978 年，4870
安娜 · 卡斯泰利 · 费里埃里，
Kartell，1985 年
可堆叠椅子

实用性强、外观如一的设计，经济实惠。

⋀ 1987, 4870
Anna Castelli Ferrieri, Kartell, 1985
Stackable chair

A practical and visually consistent
design, integrating economy of use
and production.

⋀ 1979 年，玩偶
马里奥 · 贝里尼，B&B Italia，1972 年
软垫家具系列

玩转造型设计：包裹在内的钢架给予产品强
有力的支撑，可以塑型的软垫赋予沙发柔软
的外形。

⋀ 1979, Le Bambole
Mario Bellini, B&B Italia, 1972
Series of upholstered furniture

It smartly plays with images, by concealig a
solid steel structure within the appearance
of a moldable set of cushions.

⋀ 1979 年，Maralunga 675
维科 · 玛吉斯特雷蒂，Cassina 公司自
1973 年起生产，Cassina 公司 iMaestri
系列藏品
扶手椅

标志性的可翻转靠背打造出无与伦比的灵活
性，使 Maralunga 可以满足不同需求，适
合各类场合，却又不失独特格调。

⋀ 1979, Maralunga 675
Vico Magistretti, production Cassina
since 1973, Cassina iMaestri Collection
Armchair

The flipping backrest solution provides
unique identity and flexibility, making
Maralunga suitable for different needs
and positions while remaining constantly
recognizable.

⋀ 1979 年，阿里安特
马尔科 · 扎努索，Vortice，1973 年
风扇

充满机械风的实用型工业产品，更像是图形
设计作品。

1979, Ariante
⋀ Marco Zanuso, Vortice, 1973
Fan

The interpretation of function and the
disposition of mechanical component
converge into a project almost
belonging more to graphic design tha
industrial production.

⋀ 1979 年，德尔菲娜
恩佐 · 马里，Driade，1974 年
椅子

装扮结构："德尔菲娜"椅子是人体工学轻
质结构和拉链织物坐垫的集大成者。

⋀ 1979, Delfina
Enzo Mari, Driade, 1974
Chair

Dressing the structure: the identity of the
chair comes from the ergonomics-shaped
light frame and the zip-up, textile-inspired
seating components.

⋀ 1979 年荣誉提名奖，提齐奥
理查德 · 萨珀，Artemide，1972 年
灯具

可调节平衡臂灯，既是纯粹的图形标志，也
是可调节的室内光源。

⋀ Honorable Mention 1979, Tizio
Richard Sapper, Artemide, 1972
Lamp

By working on balancing leverages, the
lamp becomes both a pure grahic sign and
an adaptable interior lighting device.

1979 年，9090
理查德 · 萨珀，Alessi，1978 年
浓缩咖啡机

极致展现"当代设计照古今"。

1979, 9090
Richard Sapper, Alessi, 1978
Espresso coffee maker

A contemporary design echoeing both
ancient and modern tradition.

1979 年，环礁 233/D
维科 · 玛吉斯特雷蒂，Oluce，1977 年
灯具

绝对几何体构成的设计，赋予这盏灯抽象的
实体感。

1979, Atollo 233/D
Vico Magistretti, Oluce, 1977
Lamp

The absolute geometries composing the
object give this lamp an abstract and
extemely physical nature at the same time.

1979 年，圆括号
阿希尔 · 卡斯蒂廖尼和皮奥 · 曼祖，
Flos，1971 年
吊灯

钢索和括号型金属管间的拉扯，凭借直觉形
成设计造型。

1979, Parentesi
Achille Castiglioni, Pio Manzù, Flos, 1971
Suspension lamp

The intuition of a physical phenomenon
– the tension between a steel wire and
a bracketed metal tube – capable of
generating form.

1979 年，Sciangai
乔纳森 · 德帕斯、多纳托 · 杜尔比诺、保罗 ·
洛马齐，Zanotta，1973 年
衣架

用最简单的构件，以最简单的方式，构建出
复杂的设计。

1979, Sciangai
Jonathan De Pas, Donato D'Urbino,
Paolo Lomazzi, Zanotta, 1973
Floor coat hanger

A complex object born from the simplest
interaction between the simplest elements.

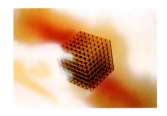

1981 年，研究
阿基米亚工作室
亚历山德罗 · 门迪尼设计的普鲁斯特扶手椅

阿基米亚工作室对设计、生产、参考资料、
历史间的关系见解独到，以此获得殊荣。

1981, Alchimia research
Studio Alchimia
Proust Armchair by Alessandro Mendini

Studio Alchimia is awarded for a
groundbreaking criticism of the
relationship between design, production,
references and history.

1987 年，托尼埃塔
恩佐 · 马里，Zanotta，1985 年
椅子

植根于设计和手工艺历史的沃土，具有原型
价值的当代设计。

1987, Tonietta
Enzo Mari, Zanotta, 1985
Chair

In giving contemporary declination to a long
history of design and craftsmanship, the
shape takes on a new archetypal value.

1984 年，电视开场
埃托雷 · 维塔莱，意大利广播电视公司
（RAI），1984 年
埃托雷 · 维塔莱提供

图形设计与复杂操作融为一体的成功尝试，
不乏受到电视行业商业化的影响。

1984, Television opening sequences
Ettore Vitale, RAI Radiotelevisione
italiana, 1984
Courtesy Ettore Vitale

A successful attempt to integrate graphic
design into complex operations and
sometimes contaminated by the commercial
aspects of the television industry.

⋀ 1994 年，卡通
路易吉 · 巴罗利，Baleri Italia，1992 年
分隔墙

用料少，可塑性强。

⋀ **1994, Cartoons**
**Luigi Baroli, Baleri Italia, 1992**
**Dividing wall**

The poverty of the material manages to produce an almost sumptuous plastic effect.

⋀ 1989 年，洛拉
阿尔贝托 · 梅达和保罗 · 里扎托，
Luceplan，1987 年
灯具

科技研究的成功典范，又不会太过张扬。

⋀ **1989, Lola**
**Alberto Meda, Paolo Rizzatto, Luceplan, 1987**
**Lamp**

A successful synthesis of technological research, with no concession to theatricality.

⋀ 1994 年，汉娜
安娜 · 卡斯泰利 · 费里埃里，
Sambonet，1992 年
餐具

功能与形式融合而成的新经典，在整体中找到个体元素的平衡。

⋀ **1994, Hannah**
**Anna Castelli Ferrieri, Sambonet, 1992**
**Cutlery service**

Reinterpretation of a classic typology where functional attention and formal care of each of the elements blends harmoniously into the balance of the whole.

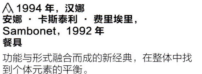

⋀ 1994 年，UniFor 企业形象设计
皮尔路易吉 · 塞里，UniFor，
1987—1994 年
企业形象设计

长期组织、管理各类艺术品和交流活动，形成始终如一的企业形象。明快的色彩和图码构成辨识度极高的企业标识，足以经受时间的考验。

⋀ **1994, UniFor corporate image**
**Pierluigi Cerri, UniFor, 1987-94**
**Corporate image**

The result of the coherent direction of a long term management process through the organizing of a wide range of artefacts and communicative events. An objective and highly recognizable identity obtained through the adoption of euphoric colour choices and a graphic code that is universal enough to promise not to age over time.

⋀ 1994 年荣誉提名奖，巴蒂斯塔
安东尼奥 · 奇特里奥和 G.O. 勒夫，
Kartell
可伸缩折叠手推车

特别适合家庭使用场景，满足各种需要，折叠收缩后节省空间。

⋀ **Honorable Mention 1994, Battista Antonio Citterio and G.O. Löw, Kartell**
**Extendable and folding trolley**

It brings into the house an empathetic object that adapts to the needs by folding itself with a pantograph system.

⋀ 1998 年，翼
詹路易吉 · 兰多尼，Rapsel，1996 年
洗手盆

形式上强调垂直感，极具创新意味，但功能性丝毫不受影响。

⋀ **1998, Wing**
**Gianluigi Landoni, Rapsel, 1996**
**Handbasin**

It is formally resolved in proportions that emphasize its verticality, giving it an innovative image without compromising functionality.

⋀ 1994 年，Cosmit 企业形象设计
马西莫 · 维涅里和维涅里协会，Cosmit，1994 年
企业形象设计

整个设计充满活力，色彩运用炉火纯青。既无冗余的图形重复，又保持了必要的庄重。不同元素"济济一堂"，生动活泼，不落窠臼。

⋀ **1994, Cosmit corporate image**
**Massimo Vignelli, Vignelli Associates, Cosmit, 1994**
**Corporate image**

Extremely dynamic and characterized by an excellent use of colour, it avoids the frequent repetitiveness in this sector of graphics without giving up a substantial seriousness. It also brings together different elements in a lively and unconventional way.

⚠ 1998 年，移动 & 翻转
卢西亚诺 · 帕加尼和安吉洛 · 佩尔韦西，
UniFor，1996 年
办公家具

确实给人以形式上完整的观感，同时又不像
一台冷冰冰的机器，看上去只是一张简洁的
办公桌。

⚠ 1998, Move e Flipper
Luciano Pagani, Angelo Perversi,
UniFor, 1996
Office furniture

In addition to communicating an
impression of formal completeness, it has
the advantage of not appearing like a
machine but a simple work table.

⚠ 2001 年，泡泡俱乐部
菲利普 · 斯塔克，Kartell，2000 年
扶手椅及沙发

用出乎意料的方式再现记忆中的样式，采用
尖端工艺生产。

⚠ 2001, Bubble Club
Philippe Starck, Kartell, 2000
Armchair and sofa

The presentation of a mnemonic form in an
ironic key adopts sophisticated techniques
in production.

⚠ 2004 年，管道
赫梭与德梅隆，Artemide，2002 年
吊灯

灵活、精致、巧妙、有趣。

⚠ 2004, Pipe
Herzog & De Meuron, Artemide, 2002
Suspension lamp

Flexible, subtle, sophisticated and playful.

⚠ 2001 年，T-Age Suit
戴尼斯 · 阿尔多 · 德鲁迪研究中心，
Dainese，2000 年
一件式摩托车骑行皮衣

既考虑到摩托骑行中的安全防护问题，又充
分展现出人体线条的美感。

⚠ 2001, T-Age Suit
Dainese Study Centre, Aldo Drudi,
Dainese, 2000
One piece motorcycling leathers

While aimed at guaranteeing safety in the
projected competitive area, [it] does not
exclude the presence and the will to also
take into consideration the expressive value
of its configuration for the human body.

⚠ 2004 年，托洛梅奥
布鲁诺·雷纳尔迪，Moco – Minotti Italia
Trading，2002 年
书柜

出人意料的设计，把地板和桌子从一摞摞的
书中解放出来，也是让人浮想翩翩的雕塑艺
术，彻底冲破了传统书柜的藩篱。

⚠ 2004, Ptolomeo
Bruno Rainaldi, Moco – Minotti Italia
Trading, 2002
Bookcase

An idea full of irony with which to free up
floors and tables from books, a suggestive
sculpture that vertically overturns the
traditional bookcase.

⚠ 2004 年，制动系统
布雷博技术部，Brembo，2003 年
碳陶瓷刹车盘和卡钳

就算不是高性能碟刹，也会是件值得现
代艺术馆收藏的雕塑品。

⚠ 2004, Impianto frenante
Brembo Technical Department,
Brembo, 2003
Carbon ceramic brake disc and
caliper

If it were not a high-performance brake,
it would be a sculpture worthy of any
modern art museum.

⋀ 2008 年，大
马克 · 萨德勒，Caimi Brevetti，2004 年
书柜

造型上线条清晰，毫不拖泥带水，关注组装过程
中的每一个细节。

⋀ **2008, Big**
**Marc Sadler, Caimi Brevetti, 2004**
**Bookcase**

The clear formal solution is combined with the
precise study of details, which facilitate the
assembly and mounting operations.

⋀ 2008 年，MT3
罗恩 · 阿拉德协会，Driade，2005 年
摇椅

正是对生产工艺长期不懈的研究，才造就了
如今的双色扭转结构。

⋀ **2008, MT3**
**Ron Arad Associates, Driade, 2005**
**Rocking armchair**

A long study into production technology has
made it possible to carry out the rotational
moulding of furnishing elements in two-
colour material.

⋀ 2011 年，色调
大卫 · 奇普菲尔德，Alessi，2009 年
餐具

具有图画意象的和谐优雅。

⋀ **2011, Tonale**
**David Chipperfield, Alessi, 2009**
**Table service**

Harmonic elegance characterized by
pictorial references.

⋀ 2011 年，螺旋桨
布赖恩 · 西罗尼，Martinelli Luce，
2009 年
台灯

轻盈的灯臂和粗壮的台座对比鲜明，所以技
术细节全部隐藏不见。

⋀ **2011, Elica**
**Brian Sironi, Martinelli Luce, 2009**
**Table lamp**

For the contrast between the lightness of
the arm and the strength of the support
accompanied by the elimination of every
visible technical detail.

⋀ 2011 年，Myto
康斯坦丁 · 格西奇，Plank，2007 年
椅子

巧妙利用塑料材质，完美兼顾结构与灵活性。

⋀ **2011, Myto**
**Konstantin Grcic, Plank, 2007**
**Chair**

To solve the problem of structure and
flexibility through an intelligent use of
plastic.

⋀ 2011 年，菲亚特 500
菲亚特集团汽车设计，菲亚特集团汽车部，
2007 年
汽车

意大利经典设计重现，毫无古早之感。

⋀ **2011, 500**
**Fiat Group Automobiles Design, Fiat**
**Group Automobiles, 2007**
**Car**

The ability to reinterpret an icon of Italian
design without nostalgic references.

⋀ 2011 年，史密斯
乔纳森 · 奥利瓦雷斯，Danese，2007 年
可移动多功能集装箱

高效、实用，功能非凡。

⋀ **2011, Smith**
**Jonathan Olivares, Danese, 2007**
**Portable multifunctional container**

Productive and constructive intelligence
accompanied by remarkable multi-
functionality.

⋀ 2011 年，Domo – XIX Biennale Dell'Artigianato Sardo Sardinia Autonomous Region，Ilisso Edizioni
书
经典再现特定环境中的工艺角色。

⋀ 2011, Domo - XIX Biennale Dell'Artigianato Sardo Sardinia Autonomous Region, Ilisso Edizioni
Book

A critical re-interpretation of the role of craftsmanship in a specific context.

⋀ 2014 年，陀螺
托马斯 · 赫斯维克，Magis，2010 年
旋转扶手椅

用新奇、有趣的方式，重新设计司空见惯的物品。

⋀ 2014, Spun
Thomas Heatherwick, Magis, 2010
Rotating armchair

Having recreated an everyday object in an ironic and entertaining way.

⋀ 2014 年，杜卡迪魔鬼 1260 s
杜卡迪设计中心 – 法布罗 · 贾南德里亚，杜卡迪，2011 年
摩托车

将赛级性能融于优雅的大众版本，不失品牌一贯的传统形象。

⋀ 2014, Ducati Diavel 1260 s
Ducati Design Center-Fabbro Gianandrea, Ducati, 2011
Motorbike

Having transferred competition level performance to an elegant mass-production model that is still consistent with the brand's traditional image.

⋀ 2014 年，Sfera
朱利奥 · 亚凯蒂和马泰奥 · 拉尼，Montini，2012 年
井盖

街道上随处可见的功能性建筑材料也有了"脑洞大开"的丰富表达。

⋀ 2014, Sfera
Giulio Iacchetti, Matteo Ragni, Montini, 2012
Manhole cover

Having interpreted a functional element of street furniture in an expressive and ironic way.

⋀ 2014 年，桑贝
恩佐 · 卡拉布雷斯和达维德 · 格罗皮，Davide Groppi，2011 年
落地灯

可在吊灯和落地灯间自由转换。

⋀ 2014, Sampei
Enzo Calabrese, Davide Groppi, Davide Groppi, 2011
Floor lamp

For its capacity to be both a suspension lamp and a floor mounted one at the same time.

⋀ 2014 年，法拉利 12 缸
弗拉维奥 · 曼佐尼 – 宾尼法利纳法拉利风格中心，法拉利，2012 年
汽车

流线型外形的空气动力学充分优化了整车性能。

⋀ 2014, F12 Berlinetta
Flavio Manzoni-Ferrari Style Centre, Pininfarina, Ferrari, 2012
Car

For its streamlined shape, whose aerodynamics optimise the car's performance.

⋀ 2014 年，平衡臂
丹尼尔 · 雷巴肯，Luceplan，2012 年
灯具

在静态与机械间创造出的诗意平衡。

⋀ 2014, Counterbalance
Daniel Rybakken, Luceplan, 2012
Lamp

Having created a poetic solution to a static and mechanical problem.

⋀ 2014 年，Inventario
贝佩 · 费内西、阿特米奥 · 克罗托 –Designwork（艺术总监）、Fascarini（发起人）、Corraini（编辑）
书刊

用具有冲击力的视觉形象、高品质的编辑设计，轻松阐释复杂的文化概念。

⋀ 2014, Inventario
Beppe Finessi, Artemio Croatto - Designwork (art director), Foscarini (promoter), Corraini (editor)
Bookzine

For the ability to summarize culturally complex concepts with a lightness of touch, illustrating them with a strong visual identity and a high quality editorial design.

⋀ 2016 年，上升
丹尼尔 · 雷巴肯，Luceplan，2013 年
带固定支杆或底座的台灯

美在简单之中，调谐实在的形体与发散的灯光。

⋀ 2016, Ascent
Daniel Rybakken, Luceplan, 2013
Table lamp with fixed hinge or with base

The simple beauty of a gesture that calibrates the intensity and spread of the light.

⋀ 2016 年，黑曜石
马里奥 · 特里马奇，Alessi，2014 年
浓缩咖啡机

传统与创新碰撞激荡，时光流逝，意式风物永恒。

⋀ 2016, Ossidiana
Mario Trimarchi, Alessi, 2014
Espresso coffee maker

Tradition and innovation in a typically Italian object that hitherto seemed unchangeable.

⋀ 2016 年，卢塞塔
埃马努埃莱 · 皮佐罗鲁索，Palomar，2013 年
磁性自行车灯

小产品，大用途，满足日常所需。

⋀ 2016, Lucetta
Emanuele Pizzolorusso, Palomar, 2013
Magnetic bicycle light

An excellent tiny product. An article that gives an intelligent answer to an objective need.

⋀ 2016 年，OK
康斯坦丁 · 格西奇，Flos，2014 年
吊灯

这盏灯，既展现出标志性产品的流变，又保有着久已有之的优雅。

⋀ 2016, OK
Konstantin Grcic, Flos, 2014
Suspension lamp

An object that presents the evolution of an icon while maintaining its traditional elegance.

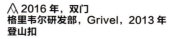

⋀ 2016 年，双门
格里韦尔研发部，Grivel，2013 年
登山扣

形式简单的专业配件，极大改善保护性能和
安全性能。

⋀ **2016, Twin Gate**
**R&D Grivel, Grivel, 2013**
**Mountaineering carabiner**

A specialised object that significantly
improves protection and safety while
respecting simplicity of form.

⋀ 2018 年，尼古拉特斯拉
法布里奇奥 · 克里萨，Elica，2016 年
带集成式抽油烟机的电磁炉

备餐似乎越来越遵循无菌原则，相关设备几
乎抽象化，由此，为设计创造留下了充分的
空间。

⋀ **2018, Nikolatesla**
**Fabrizio Crisà, Elica, 2016**
**Induction hob with integrated hood**

The functions of food preparation seem
to be increasingly dictated by the logic of
aseptic, almost abstract instruments, in
which communication of the function leaves
ample room for subjective interpretation.

⋀ 2018 年，折纸
阿尔贝托 · 梅达，Tubes Radiatori，
2016 年
散热器

总能成为所在空间焦点的实用性物品，使用
者可以随心调整。

⋀ **2018, Origami**
**Alberto Meda, Tubes Radiatori, 2016**
**Radiator**

A functional object that takes centre stage
in the space which it occupies and which
adapts to the user to whom it relates.

⋀ 2017 年国际奖，Bolide TT
皮纳雷洛实验室，Cicli Pinarello，2016 年
计时赛自行车

计时赛自行车的新基准。设计与功能：空气动
力学、轻量化、美学。

⋀ **International Award 2017, Bolide TT**
**Pinarello Lab, Cicli Pinarello, 2016**
**Time trial bicycle**

A new benchmark for time trial bicycles.
Design and functionality: aerodynamics,
lightweight, aesthetics.

⋀ 2018 年，探索吊灯
埃内斯托 · 吉斯蒙迪，Artemide，2016
年
照明设备

关闭电源后，细细的金属圈在半空中圈出一
片寂静。打开电源，发出强烈光线的金属盘
诠释出这件产品的真谛。

⋀ **2018, Discovery Sospensione**
**Ernesto Gismondi, Artemide, 2016**
**Llighting device**

When turned off it is a slender metal
ring that describes a silent emptiness
suspended in space. Turned on, it suddenly
becomes an intensely luminescent disc
creating a surprise that is the true soul of
this product.

⋀ **2020 年，E-Lounge**
**安东尼奥 · 兰奇洛建筑事务所，**
**Repower，2017 年**
**充电长椅**

不同设计元素的综合产物，涵盖数字、共享经济、社群文化、城市家具和互联互通诸多方面。

⋀ **2020, E-Lounge**
**Antonio Lanzillo & Partners, Repower,**
**2017**
**Electrified service bench**

A new type of product able to combine different design aspects: Digital, sharing economy, neighbourhood culture, urban furniture and connection.

⋀ **2020 年，航空**
**保罗 · 卡塔内奥和克劳斯 · 菲奥里诺 –Momodesign，Momodesign，2018 年**
**摩托车头盔**

面罩从盔顶一泻而下，铺陈出流动的空气动力美学。

⋀ **2020, Aero**
**Paolo Cattaneo, Klaus Fiorino –Momodesign, Momodesign, 2018**
**Motorcycling helmet**

The visor sits flush with the shell and generates a fluid and aerodynamic aesthetic.

⋀ **2020 年终身成就奖，拱**
**阿希尔 · 卡斯蒂廖尼与皮耶尔 · 贾科莫 · 卡斯蒂廖尼，Flos，1962 年**
**灯具**

照明领域的原创经典，已成为全球公认的意式设计典范。

⋀ **Career Award 2020, Arco**
**Achille and Pier Giacomo Castiglioni,**
**Flos, 1962**
**Lamp**

Typological innovation in the lighting sector, over time it has become an icon for Italian design worldwide.

⋀ **2020 年，汉内斯**
**IIT 意大利理工学院和 INAIL 假肢中心 Ddps 工作室洛伦佐 · 德 · 巴托洛梅斯、加布里埃莱 · 迪亚曼蒂、菲利波 · 波利，2017 年**
**假肢**

科技与美学共筑身心健康。

⋀ **2020, Hannes**
**Lorenzo De Bartolomeis, Gabriele Diamanti,**
**Filippo Poli - Ddpstudio, IIT Istituto Italiano di**
**Tecnologia, INAIL Centro Protesi, 2017**
**Prostheses**

Technology and aesthetics come together to help overcome psychological distress and physical shortcomings.

⋀ **2020 年，D-Heart**
**意大利设计集团，D-Heart，2017 年**
**智能手机心电图系统**

医学科技逐渐为人熟知、融入日常生活。便于使用，可以远程监控患者情况。

⋀ **2020, D-Heart**
**Design Group Italia, D-Heart, 2017**
**ECG system for smartphones**

Medical technology made familiar and transposed into daily life. Completely user friendly while allowing the patient to be followed from a distance.

⋀ **2020 年荣誉提名奖，Agritube**
**帕多瓦大学 DAFNAE 部，帕多瓦大学全球影响力网络，2018 年**

简化的地面水培农业系统，在循环经济大背景下达成农业体系的经济可持续发展目标。

⋀ **Honorable Mention 2020, Agritube**
**Università di Padova, Dipartimento DAFNAE,**
**Glocal Impact Network, 2018**

A simplified system of above-ground hydroponic agriculture that has as its main goal the economic sustainability of the agricultural system, within a circular economy perspective.

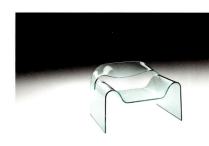

## 2022 年终身成就奖，幽灵
### 奇尼 · 博埃里和托姆 · 卡塔扬吉，Fiam Italia，1987 年
### 扶手椅

奇尼 · 博埃里和托姆 · 卡塔扬吉设计的幽灵扶手椅是技术试验与形式研究的完美结合，是从使用者角度出发，对产品功能去实物化的一大尝试，使用者由此成为使用空间内无可争议的主角。

## Career Award 2022, Ghost
### Cini Boeri, Tomu Katayanagi, Fiam Italia, 1987
### Armchair

A perfect synthesis of technological experimentation and formal research, the Ghost armchair designed by Cini Boeri and Tomu Katayanagi represents the desire to dematerialize the perception of function in favour of the user, who thus becomes the undisputed protagonist of the space."

## 2022 年，Contro l' Oggetto. Conversazioni sul Design
### 埃马努埃莱 · 奎因兹，Quodlibet，2020 年
### 书

有关 21 世纪设计本质的一连串对话。质疑设计概念和功能的最新力作。

## 2022, Contro l'Oggetto. Conversazioni sul Design
### Emanuele Quinz, Quodlibet, 2020
### Book

A series of conversations on the nature of 21st century design. An updated and contemporary book that questions the concept and function of design.

## 2022 年，LAMBROgio e LAMBROgino
### 牧尾莲池公司，Repower，2019 年
### 电动车

用新的都市电动车重新定义传统的轻型交通工具，载人载物两相宜。

## 2022, LAMBROgio e LAMBROgino
### Makio Hasuike & Co., Repower, 2019
### Electric cycle-vehicles

A brilliant redefinition of traditional lightweight vehicles for new urban electric mobility aimed at the transport of people and the delivery of goods.

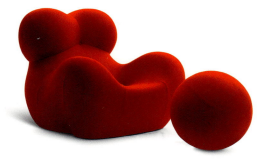

## 2022 年终身成就奖，"向上" 5
### 加埃塔诺 · 佩斯切，B&B Italia，1969 年
### 扶手椅

加埃塔诺 · 佩斯切设计的"向上"扶手椅是技术试验与形式研究的完美结合，是对女性形体与各类社会文化语境间复杂关系的诗意观察，体现出设计的责任感。

## Career Award 2022, Up 5
### Gaetano Pesce, B&B Italia, 1969
### Armchair

The perfect synthesis of technological experimentation and formal research, the Up armchair designed by Gaetano Pesce is a poetic witness to the complex relationship between the female figure and different socio-cultural contexts and at the same time a product which has always represented a vision of civic commitment to design.

## 2022 年荣誉提名奖，Q/Wood
### 菲利普 · 斯塔克，Kartell，2020 年
### 用于生产和木工系统的材料

使用专利技术，利用专门模具，把木板弯成型到满意的角度，形成凹凸有致的座椅。

## Honorable Mention 2022
### Q/Wood Philippe Starck, Kartell, 2020
### Materials for production and woodworking system

Using patented technology, the wood is worked with a mould designed to bring the curvature of the panel to the very edges, creating a seat with sinuous roundness.

# 意大利风格：
# 设计趋势

迈特 · 加西亚 · 桑奇斯

# Italian Way:
# Design Trends

Maite García Sanchis

纵观金圆规奖 70 年历史，设计师、生产商和设计作品凝结为一个整体，描摹出意式设计的全景，也勾勒出设计界的挑战和趋势。这场为中国公众策划呈现的"引领潮流七十年——意大利金圆规奖经典设计作品展"，不仅要展示获奖作品，更要传达这些作品在设计领域内的重要意义；它们是过去的里程碑，也是设计界未来发展的动力源泉。鉴于此，展览以二元方式展开。一方面，沿着"时间线"铺陈当代设计中的节点性作品，展示文化变革在设计领域内的标志性转折点。展品配有评委会的获奖作品说明节选，简要描述、评论每件展品，同时介绍 1954 年至 2022 年历届金圆规奖的文化、社会背景。另一方面，我们又摆脱了时间走向的束缚，围绕不同主题布置了若干分展区，展示的获奖作品不仅自身具有突出价值，而且也是设计趋势的弄潮儿，超越了历史、类型、功能、美学的分类标准。分展区中的获奖作品、设计、项目打乱了时代顺序，聚集在不同主题之下，描绘出推动意式设计发展、呈现其历史演变与应对当下挑战的七大主题。此次的策展方式抛开传统的分类方法，以整体性更强的情境式布展开启新洞察，引入新角度，有助于更好地理解当代社会中的设计，揭示金圆规奖这一媒介的必要性。

展览把展品置于历史、社会、文化的大背景之下，揭示新的作品含义与重要意义，强调设计在功能性和美学特性之外的影响和启迪。因此，参展的获奖作品或设计按照主题分类，避免使用枯燥的文字描述和置景，确保呈现给公众的图纸、视频和原型作品有助于再现最终设计所处的时代潮流。我们希望勾画出与过去和当下设计交织缠绕的文化脉络，这些脉络彰显着意式设计的文化、社会、产业属性。最初的主题岛台构想最终演化为主题房间，但并没有采用复原 20 世纪初房间摆设的方法，而是使用多种布展设计，贴合分展区的主题趋势。整个展览在布展和寓意上都是开放性的：既有包含不同设计版本的完整系列，也有单独一件展品；既可以是生产中用到的工具、材料，也会是图形设计和声音。这一策展方法恰恰体现了设计的多面性，涵盖了产业挑战、文化内涵、日常影响建构等方方面面。

入选本次展览的设计趋势充分贴合意式设计和金圆规奖。尽管金圆规奖针对生产商，而非设计师，但在不乏阿德里亚诺 · 奥利韦蒂、朱塞佩 · 布里翁等名人的工业、文化背景下，也有必要为设计师和设计业创造合作空间，凸显奖项能够取得成功，得益于其本身汲取传统、拥抱技术创新的强大能力，以及设计对日常生活的意义。在"工艺与工业"展区，有一张 20 世纪 50 年代 Cassina 工厂的照片，由吉奥 · 蓬蒂设计的著名 699 超轻椅悬在半空，一群工人和参观者正在围观；照片旁边则是拆解的 Laleggera 椅子——由里卡尔多 · 布卢默设计、Alias 生产，获 1998 年金圆规奖。这两把椅子的研究、设计一脉相承，穿越 40 年时光，又在今日的展览上相遇，传承发展的力量。在同一展区中还有马尔科花瓶、Tite 灯、Klipper 隔音系统，清晰展现技术、材料、设计流程内的潮涌——传统与创新联手推进设计，匠人精神与最新科技相互融合。我们从材料、科技角度看待设计业的发展，也展示对设计的理解发生了何种变化。"家居生活"展区展示了从标志性单品设计到产品体系设计的演变。里卡尔多 · 达利西为 Alessi 品牌设计的那不勒斯咖啡机原型（1981 年获奖）说明同一件物品可以担任不同的角色。和咖啡机原型共处一室的还有其他产品系列，例如：米凯莱 · 德卢基为 Artemide 品牌设计的托洛梅奥系列灯具，是单盏灯具向当代照明系统发展的代表性产品；奇尼 · 博埃里的"条纹

家族"使用可以自由组合的坐具模块完美体现系列产品的概念；迈克尔 · 阿纳斯塔夏季斯为 Flos 品牌设计的"安排"则属于开放式照明组合设备。

"住所无定式"展区既是对意式设计的展示，也是对文化里程碑重要性的表达。这个小空间无疑是在向 1972 年现代艺术博物馆的安姆巴斯"意大利：新家庭景观"展致敬，探寻向外延伸的居家生活概念；扩展的家庭空间模糊了室内室外的边界，始终是设计研究的一大焦点。展品在时间上可以追溯到 1963 年乔 · 科隆博设计的迷你厨房（获 2022 年金圆规终身成就奖），跨越数十年到 2019 年阿德里亚诺设计公司为 Fabita 公司设计的"秩序"电磁炉。对家用产品的类型分类法的深入探究，引发对资产阶级生活空间的微妙质疑。如果从带有朋克色彩的视角来看，熊猫车（1981 年获金圆规奖）灵活多变、性价比高的内部空间，就是资产阶级生活空间的体现。其商业广告通过视频和海报主打宣传这一老少咸宜、包容一切的家用空间。布展过程中，我们选择恰当的媒体手段展示档案素材，增强展览主题的表现力，营造展品所处的历史环境，与展厅彩色墙壁上的说明和配图文字相得益彰。很多时候，一项设计面世之后经历了漫长的时间，往往会导致忽略掉这一设计为解决当时所面临的问题而提供的创新方案。而菲亚特公司对熊猫车的广告营销却恰恰贴合本次展览的策展理念，不只要解释说明作品本身，更要凸显设计原本的内在精髓和意图。

除了广告资料之外，历史出版物也有助于建构设计作品或项目的背景，把展品置于特定主题之下。跟随一系列精心挑选的文章标题，参观者可以把获奖作品连缀到意式设计的主线上，并将其与特定时代背景关联起来。"发明与创造"展区就是在历史出版物中的一次徜徉。展品中有 1961 年的《L'Europeo》杂志，上面刊登了吉诺 · 科隆比尼为 Kartell 公司设计的 KS1171 塑料沥水架照片，同时刊出的还有 BBPR 和卡斯蒂廖尼兄弟的作品。穆纳里的橡胶玩具"仔仔"也曾出现在多本期刊中，有照片、图纸，甚至是 X 光片的相关分析材料，也一并纳入了本次展览，与公众见面。同时展出的还有"发光者"和"希望"灯具、"小章鱼"叉匙，它们都是设计天赋与原型碰撞的产物：卡斯蒂廖尼兄弟的"发光者"是对彼得罗 · 基耶萨早期版本的更新；"希望"脱胎于枝形吊灯；"小章鱼"将餐叉与餐勺合二为一。"转折点"展区的展品更是改变了游戏规则，重新定义了产品的分类属性，开拓了设计门类新领域。1967 年获奖的"蟋蟀"电话是典型个人通讯工具，造型紧凑，适合单手握持；皮耶罗 · 加蒂、切萨雷 · 保利尼、佛朗哥 · 泰奥多罗在 1968 年为 Zanotta 品牌设计的豆袋沙发，是对传统椅子的颠覆；60 年代初，马尔科 · 扎努索和理查德 · 萨珀为 Brionvega 品牌设计的"多尼"便携式电视，让超级互联就在眼前。不过，档案资料是以视频的形式呈现，展台上摆放着获奖作品实物以及原型概要；通过与西比克团队的合作，以五大明显对比展示意式设计对全球设计转型、发展的重大影响。

与此同时，"传播的力量"展区布满了海报、说明文字、图像，介绍平面设计在打造标志性产品中发挥的作用，以及对于 Olivetti、Campari、Solari 和 Artemide 等产品生产商的重要意义。这种溢出设计本身的力量塑造着时代的形象，也打造着地点的风貌。在 Solari 翻牌式显示器的噼啪声中，瞬间仿佛置身于车站、机场的人流汹涌；Olivetti 公司的"情人"和"字母 22"打字机广告宣示了大众传媒新

时代的到来；Campari（金巴利酒）海报以标志性的金巴利红锥形瓶渲染着意式生活风情；在马吉斯特雷蒂的"月食"台灯上，能看到平面设计对产品设计的影响。有了声音和颇具冲击力的平面设计元素的加持，"传播的力量"展区华丽登场。

最后要谈一下"时间线"上的第一个分展区"公共服务"。这个展区包括 60 年代初的两项设计和 2020—2022 年间的三项获奖设计，尽管时间跨越半个多世纪，这五项设计都被视为是新式设计的典型代表。我们希望能把以人为本的设计精髓传承下去。阿尔比尼、赫尔格、努尔达在 1963 年设计的米兰地铁，是米兰市内设计协调统一的最大公共空间；"心灵食粮"社会公益项目解决食物浪费问题，把优质设计带入社区厨房。两个项目均以真诚为底色，在公益领域相互呼应。扎努索和萨珀在 1964 年设计的 K1340 儿童椅，其设计理念也在 2020 年荣誉提名奖作品 Hackability 模型中有了回响——设计，就是为了满足人的需求而生。此外，Isinnova 公司的 3D 打印 Easy-Covid 19 医用呼吸机由浮潜面罩改造而来，在新冠疫情暴发后确实能够挽救生命。这一开源设计表明设计的未来趋势将凸显适应性设计、开源知识、普遍性全球目标。

我们沿着"时间线"的线性结构，铺陈展开七大主题展览，洞察意式设计的精髓，揭示金圆规奖的本源，呈现历届藏品的不凡。正因为此，我们才回溯 70 年，以当代眼光、带着当下的紧迫，重新发现以往的作品、项目、服务、空间，审视那些将会塑造未来的共通趋势。

Over its seventy-year history, Compasso d'Oro has built a community of designers and producers and a collection of items which offer a complete depiction of the character of Italian design and of the challenges and trends of design at large. When presenting this collection to the Chinese public at the exhibition *Compasso d'Oro Award: Seventy Years of Leading Italian Design Trends*, the curatorial approach aims not only to showcase the individual icons that have received the award but also to convey their broader significance in the design scene, as milestones of the past and as dynamising tools of the ever-evolving design scene. This idea was translated in a bipartisan layout of the exhibition. On one hand the 'Timeline' introduces the visitor to a chronological sequence of objects that have shaped the evolution of modern design with iconic 'Turning points' for cultural evolution in the design realm. This display is accompanied by excerpts of the motivations given by the juries of the Compasso d'Oro Award which not only give a critical description of the objects but also serve as testimony of the cultural and social context of each period, ranging from the first edition of the award in 1954 until the latest in 2022. On the other hand, we decided to present a series of thematic insights, independent from the chronological display, which put into value the role of many of the awarded items not only for themselves but as representatives of a series of design trends which transcend historical, typological, functional or esthetical classifications. In these insights awarded objects, projects or initiatives of different periods are presented together in order to illustrate seven driving themes of Italian design which have a significant role both in its historical evolution and in addressing contemporary challenges. By transcending conventional categorisations and embracing a more holistic and contextual approach, it becomes possible to reveal new insights and perspectives that contribute to a more comprehensive understanding of the role of design within contemporary society while enhancing the importance of Compasso d'Oro as a necessary medium to come up with these groupings.

By presenting a curated narrative that contextualises design objects within the broader historical, social, and cultural framework, it becomes possible to reveal new layers of meaning and significance. This approach emphasises the importance of considering the broader implications and impacts of design beyond their immediate functional or aesthetic attributes. In this occasion a further step was proposed in order to present a series of items or services awarded with the Compasso d'Oro Award under the light of a certain aspect, avoiding a thorough philological description and contextualisation, limiting the drawings, videos, or prototypes presented to the public to those which help to place the final design in a precise current. That was our main interest, to highlight a series of cultural threads that intertwine through historical and present designs and speak very clearly about the specificities of Italian design as a cultural, social and industrial reality. The awards are opened up, both metaphorically and physically, in varied ways. Sometimes they present a full collection of variations, while at other times, they showcase a single unique example. They may also highlight different aspects such as the tools and materials used in production, or focus on graphic design and sound. This approach embodies the multifaceted nature of design, encompassing industrial challenges, cultural implications, and the construction of common everyday imagery.

The selected trends speak both about Italian design and about the Compasso d'Oro Award. As an award that is handed to the producer, rather than the designer, within an industrial and cultural landscape that has seen prominent figures like Adriano Olivetti and Giuseppe Brion, it becomes essential to allocate space to acknowledge the collaborative efforts between designers and the design industry. It underscores the significant responsibility of its success in its capacity to learn from tradition while embracing technical innovation and the evolving significance of design in everyday life. In Crafts and Industry an image of 699 – the weight defying chair by Gio Ponti which anticipated the renown Superleggera – up in the air, surrounded by a group of workers and visitors at the Cassina factory in the 50's, encounters a dismantled Laleggera chair by Riccardo Blumer (manufactured by Alias, awarded in 1998), establishing a dialogue between two items which sail across the same research 40 years apart. Continuation and evolution. In the same room a Marco vase, a Tite lamp and a Klipper acoustic panel explicit the varied ways in which tradition and innovation work together to push forward the frontiers of design, grafting the crafting heritage with the latest technological advances, whether this may happen through techniques, materials or design processes. Our interest was also to share the evolution of the design industry not only from the material and technological point of view but from the changes it introduces in our way of understanding design. The evolution from icon design to system design is explored in the exhibition in 'Families', a room where the iconic prototypes by Riccardo Dalisi for his Neapolitan coffeepot for Alessi (awarded in 1981) are showcased as the multiple possible characters of an object and share the space with different interpretations of a family of products: Tolomeo, the family of lamps designed by Michele de Lucchi for Artemide that represents the shift between lamp icons and contemporary lighting systems; La famiglia degli Strips (The Strips Family, by Cini Boeri) which presented the family concept through a wide range of seating modules that can be combined; and Arrangements (by Michael Anastassiades for Flos), an open lighting composition.

The way in which Italian design has been exhibited and communicated and the importance of certain cultural milestones is also acknowledged, particularly 'Limitless Domesticity'. This is a room which undoubtedly pays homage to Ambasz's exhibition *Italy: The New Domestic Landscape* (MoMA, 1972), by exploring the idea of an expanded domesticity – which puts in crisis its limits inside and outside the home – as a persistent research topic, showcasing from Joe Colombo's Minikitchen of 1963 (awarded with a Compasso d'Oro Career Award in 2022) up to the Ordine induction hobs designed in 2019 by Adriano Design for Fabita). The typological classification of domestic pieces is interrogated, introducing a subtle questioning of the bourgeoise living spaces which in an almost punk way can be considered the alter ego of the flexible and affordable interiors of Panda (awarded in 1981), a vehicle here presented as a domestic space open for all and for everything through commercial advertising videos and posters. The choice of the media in which we present archival materials to the public serves as a means to bolster the thematic scope of the exhibition and provide context, complementing the statements and captions on the coloured walls of the rooms. In many cases the journey of a design after its launching eclipses some of the innovative aspects that it tackled at the time. In the case of Panda, Fiat's advertising campaign was very useful to present our curatorial approach not as an interpretation of an object but as the underlining of an intention that was originally in the design's genetics.

In addition to advertising documents, vintage publications help building context and framing an object or project within a given theme. Following the headlines of a series of selected articles not only helps the visitor in grasping the connection of the awarded designs to a thread of Italian design but also enables them to recognise its relevance within a specific time frame. This happens in 'Invention', presenting Gino Colombini's work with plastics through Kartell's KS1171 dish drainer published on an issue of L'Europeo of 1961 – besides the members of BBPR or The Castiglioni brothers –, or Munari's Zizi rubber toy in multiple journals analysed through photographs, drawings and even an x-ray. These two objects stand besides Luminator and Hope lamps and Moscardino fork, where genius meets the renewal of an archetype: the standing indirect lamp – with declared inspiration from Pietro Chiesa –, the chandelier, or a spoon and a fork. This idea is furthermore explored in the room entitled 'Turning Point' which portrays a group of objects which changed the rules of the game, redefining the characteristics of entire categories and opening new ones. From Grillo, a telephone awarded in 1967 which speaks about personal interaction with communication devices and the compression of the telephone in one piece which fits into your hand, to the complete deconstruction of the chair with the pouf Sacco (by

Piero Gatti, Cesare Paolini and Franco Teodoro for Zanotta in 1968), or the introduction to hyper-connection through Doney, a portable television set of the early 60's by Marco Zanuso and Richard Sapper for Brionvega. As an exception, archival materials are shown on screen while the scene is dominated by the awarded objects and an outline of their archetypes, in an effort, together with the team of Cibic Workshop, to present five contrasting comparisons that would self-explain the reach of Italian design in redefining and pushing forward design at large.

In an opposing strategy, The Power of Communication room is packed with posters, statements and images which introduce the significance of graphic design in building an icon and the relevancy it had for the producers on show: Olivetti, Campari, Solari and Artemide. It also speaks of how this communicative strength overflows the design itself, shaping the imagery of a certain era or a certain place. The flicking sound of Solari split-flap displays takes us immediately to the lobby of a train station or an airport; the advertising campaigns of Valentine or Lettera 22 by Olivetti introduced a new society in mass communication; the posters by Campari or its bright red cone-shaped bottle carry with them the imagery of Italian lifestyle; while the shape of Eclisse the lamp by Magistretti speaks of the influence of graphic design on product design. Introducing sound and powerful graphic elements transformed this thematic room into a piece of a vibrating realm.

The last consideration goes to the first insight that breaks the 'Timeline' of objects at Bund 18: Commons. This room features two designs of the early 60's and three designs awarded between 2020 and 2022 which reckon new forms of design. In spite of the significant time gap the scope is shared among the five awards. Our interest was to put into value this continuation which presents a side of design as a discipline at the service of the user and of communities. The genuine design approach of the Milanese underground (Albini, Helg and Noorda, 1963) – the largest public space with design unity of the city , resonates in Food for Soul, a social welfare project which tackles food waste and brings quality design to community kitchens;  while the user centred approach in the design of the K1340 children's chair by Zanuso and Sapper for Kartell in 1964 echoes in Hackability, a collective initiative which adapts existing designs to people with special needs, awarded with an Honourable Mention in 2020. Among these four awards a 3-D printer produces Easy-Covid19 medical valves, an open-source design by Isinnova to transform diving masks into medical breathing devices – which turned to be lifesaving during the first epidemic of Covid 19, making a stance about adaptive design, open-source knowledge and common global goals for the future of design.

In conjunction with the linear structure of the 'Timeline', our aim is to present the seven thematic rooms not only as insights into Italian design but also into the Compasso d'Oro Award and its collection as extraordinary testimonials. These enable us to dive into seven decades of design, rediscover objects, projects, services, and spaces through the lens of contemporary urgencies and see into the shared trends that will shape the future.

公共服务

Commons

设计师、制造商、用户"三位一体"，营造舒适空间，简化现有程序，在持续变化的社会中发现新需求，这是设计专业服务公众的关键。在本展区，您将看到近几十年设计师如何以社会公益为导向，匠心独运，打破常规，创新开发了融入米兰民众日常生活的城市地下空间，在设计童椅时不再受限于将成人座椅缩小的老套，让童椅真正适合儿童使用。而对于食物链、社会不平等现象，以及开源共享知识等公共事务中不同机会和趋势的把握，激励着设计师顺着趋势，探寻未来设计的新路径。

Designer, producer, user. This triad is the key to design intended as a discipline which is at the service of the community, creating comfortable spaces, simplifying existing processes, or recognising new demands from a society in continuous change. The interest to go beyond the obvious, and respond with new forms to the development of a new subterranean urban space meant to become the daily landscape of the Milanese and to the invention of a children's chair which exceeded from the miniaturisation of an existing seat, continues in the near past with socially driven projects that look over established protocols and contemporary dynamics. The identification of the trends in the food chain, in social inequalities and the possibilities of open-source shared knowledge give an inspiring clue of the future destiny of design trends.

孩子爬上搭叠的 K1340 儿童椅。
Climbing a composition of stacked K1340 chairs.

ADI 设计博物馆档案
Archivio ADI Design Museum

KS1340 儿童椅宣传册内页，由米凯莱 · 波尔温恰利负责艺术指导。
Extract from the brochure of the KS1340 chair, art direction by Michele Porvinciali.

Kartell 博物馆档案
Archivio Museo Kartell

画有多重投影的技术图纸，1959—1964 年。
Technical drawing with overlapping projections, 1959-1964.

巴莱尔纳当代档案馆，马尔科 · 扎努索基金
Balerna, Archivio del Moderno, Fondo Marco Zanuso

孩子在用 K1340 儿童椅搭叠的滑梯上玩耍，1964 年。
Children playing on a composition of stacked K1340 chairs, 1964.

Kartell 博物馆档案
Archivio Museo Kartell

孩子在搭叠的 K1340 儿童椅上玩耍，1964 年。
Children playing on a composition of stacked K1340 chairs, 1964.

Kartell 博物馆档案
Archivio Museo Kartell

## ⋀ 1964 年，K 1340

### 马尔科 · 扎努索和理查德 · 萨珀
### Kartell 公司 1959—1964 年

### 儿童椅

想象一下，一张简简单单的座椅，从孩童身上汲取灵感，竟摇身一变，成为了色彩纯粹、几何造型简洁的玩具。

这款儿童椅凭借可自由组装的模块化设计，能够让儿童发挥天性与本能，用小手搭建出他们心中构想的奇妙世界。

"这款儿童椅可以让孩子尽情、安全地玩耍。他们能轻松搬动这把椅子，但是又不够力气扔来扔去。椅子扛摔耐造、能水洗、便宜、无噪音。"

"观察孩子与物体之间的互动真的非常棒，我受益良多。孩子的创造力天马行空，全凭一流的直觉。" Marco Zanuso

## ⋀ 1964, K 1340

### Marco Zanuso and Richard Sapper
### Kartell, 1959-1964

### Children's Chair

What could be a simple seat that is transformed into a real toy of a simple geometry of primary colours inspired by the figure of the child.

Modular and assemblable, it is capable of generating architectures and landscapes starting from the children's gestures and their intuitive relationship with the object.

"A chair for children with which they could play, but not get hurt; light enough to be carried by children, but not thrown; indestructible, washable, not noisy and cheap."

"Observing the relationships of children with objects is fantastic, it taught me a lot. Children have an unbridled creativity full of great intuitions." Marco Zanuso

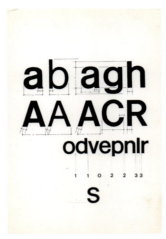

地铁标识系统设计图：字母表、字体样式、站台标牌、信息板细节、线描图，1962—1963 年。
Metro signage drawings: alphabet, studies of characters, a study of the visualisation of platform signs, details of the information boards and a study of line drawings, 1962-63.

佛朗哥·阿尔比尼基金会
Fondazione Franco Albini

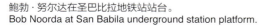

鲍勃·努尔达在圣巴比拉地铁站站台。
Bob Noorda at San Babila underground station platform.

佛朗哥·阿尔比尼基金会
Fondazione Franco Albini

地铁站内景和外景的旧照。
Historical photographs of the exterior of the subway.

ATM 历史档案
Archivio Storico ATM

大教堂站站台，米兰地铁，1964 年。
Duomo station platform, Milan Underground, 1964.

摄影：卡洛·奥尔西，卡洛·奥尔西档案
Photo by Carlo Orsi, Archivio Carlo Orsi

## ⋀ 1964 年

### 米兰地铁的标识和内部设计

### 佛朗哥 · 阿尔比尼、弗兰卡 · 赫尔格，与安东尼奥 · 皮瓦（内部设计）和鲍勃 · 努尔达（标识）合作

### 米兰地铁（M.M.）项目，1964 年

建筑 / 平面设计提案深入研究标识系统、标识间层级关系、标识位置、站内设施之间的协调统一，直观描述交通环境。

米兰地铁是米兰市内设计、建造的最大公共空间：迥然各异的地点通过沿 12 公里长的隧道分布的 21 个地铁站相互联通。该项目注重乘客体验，也是城市形象的转折点。

项目在对外的形象设计上保持一以贯之的风格：打造舒适出行方式，展现地铁这一新型公共交通方式的时代进步。

采用包含现代平面设计和最简标识系统的创新设计方案，准确勾勒城市形象，预示着这种新式美学进一步融入城市肌理。

## ⋀ 1964 Setting up and signage, Metropolitana Milanese

### Franco Albini, Franca Helg, with the collaboration of Antonio Piva (setting up), Bob Noorda (signage)

### Metropolitana Milanese M.M., 1964

The architectural and graphic proposal aims at directly describing the environment through communication through an in-depth study of the set of signals, their hierarchical relationships and their location as well as the unity of materials.

The Milan Subway became the largest public space conceived within the city: twenty-one stations along twelve kilometres of tunnels that connected very different places through a shared and common project, focused on user experience and which marked a Turning Point in terms of urban imagery.

The project's consistent character in the communication field proved to be for the traveller's comfort and the recognition of the values of progress and modernity associated with the arrival of this new means of public transport.

The choice of adopting an innovative design that gave an identity to the city's landscape through modern graphics and minimal signage presaged a step forward in confirming this new aesthetic within the urban context.

盎博罗削食堂，米兰。徽标设计草图。Origoni Steiner
联合建筑师工作室，2015 年。
Refettorio Ambrosiano, Milan. Logo design, sketches.
Origoni Steiner, 2015.

"心灵食粮"项目，"心灵食粮"组织，2015 年
Food for Soul, a project, Food for Soul, 2015

盎博罗削食堂，米兰，2015 年。
Reffettorio Ambrosiano, Milan, 2015.

盎博罗削慈善机构和"心灵食粮"项目
Caritas Ambrosiana. Courtesy Food for Soul

## ⋀ 2020 年，"心灵食粮"项目
## "心灵食粮"组织，2015 年
### 社会福利项目

每年，全球三分之一的食物惨遭浪费。

"心灵食粮"项目借助设计的力量，将食物浪费链与社区厨房相连。知名主厨、艺术家与设计师响应号召，参与到项目中来，共同营造重要空间，体现出重新思考食物链及其社会影响的重要性，处理资源浪费、稀缺与再分配的问题。

大厨们携手合作，是为了"证明回收利用的食物，无论是过熟的、碰伤的，还是过期的，还有原本会被扔掉的厨余边角料，都不仅能吃，甚至还十分美味"。

## ⋀ 2020, Food for Soul
## Food for Soul, 2015
### Social welfare project

One third of all the food produced globally is wasted every year.

Food for Soul connects the chain of food waste with community kitchens through design. Renown chefs, artists and designers are called to sum up to the project in order to create significant places that symbolise the importance of rethinking the food chain and its social impacts, dealing with waste, scarcity and redistribution of resources.

The collaboration of renown chefs aimed "to prove that salvaged food, overripe or bruised and beyond expiration dates, as well as scraps and trimmings that otherwise would be thrown away, were not only edible, but even delicious."

在"心灵食粮"食堂中节约食物。
Saving exceeding food at the *Food for* Soul Reffettorios.

照片出处：斯蒂芬妮·比托、伊曼纽尔·科伦坡、西尔维娅·科蒂切利、安吉洛·达尔·博、谢汉·汉维拉奇、西蒙·欧文、尼古拉斯·罗切特、保罗·萨利亚。盎博罗削慈善机构和"心灵食粮"项目
Photographs by Stephanie Biteau, Emanuele Colombo, Silvia Corticelli, Angelo Dal Bo, Shehan Hanwellage, Simon Owen, Nicholas Rochette, Paolo Saglia. Courtesy Caritas Ambrosiana and Food for Soul. Courtesy Food for Soul

世界各地"心灵食粮"空间开展的社区活动。
Communities in action in the *Food for* Soul venues around the World.

照片出处：斯蒂芬妮·比托、伊曼纽尔·科伦坡、西尔维娅·科蒂切利、安吉洛·达尔·博、谢汉·汉维拉奇、西蒙·欧文、尼古拉斯·罗切特、保罗·萨利亚。盎博罗削慈善机构和"心灵食粮"项目
Photographs by Stephanie Biteau, Emanuele Colombo, Silvia Corticelli, Angelo Dal Bo, Shehan Hanwellage, Simon Owen, Nicholas Rochette, Paolo Saglia. Courtesy Caritas Ambrosiana and Food for Soul. Courtesy Food for Soul

呼吸机阀门 3D 模型。
3D model of the valve.

Isinnova 公司档案
Archivio Isinnova

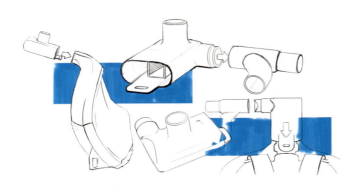

呼吸机阀门接头的草图。
Scketch of the valve's connectors.

Isinnova 公司档案
Archivio Isinnova

配有呼吸阀的医用呼吸面罩。
Mask with valve connected to medical respiration apparatus.

Isinnova 公司档案
Archivio Isinnova

## 2022 年 Easy-Covid 19
### 克里斯蒂安 · 弗拉卡西、亚历山德罗 · 罗马约利
### Isinnova 公司，2019 年
### 医用呼吸机

使用非专业 3D 打印机在短时间内就能制作的一小块塑料模型，可以将现有的标准商用浮潜面罩改造为能够挽救生命的急救医疗设备。

便捷制造是该项目的一大决定性驱动因素，这一设计也因此在 2020 年新冠疫情肆虐之时，在全世界广为流传。

该医用呼吸机的阀门由夏洛特阀和戴夫阀组成，将面罩与供氧装置相连，保证患者呼吸，同时由于该系统密封性良好，可确保适当的充气压力。

## 2022 Easy-Covid19
### Cristian Fracassi, Alessandro Romaioli
### Isinnova, 2019
### Medical valve

A small piece of plastic that can be built in short time with an amateur 3D printer shifts the use of an existing standard commercial snorkelling mask to an emergency medical device capable of saving lives.

The ease of fabrication was one of the decisive driving factors for the project, which had, during the peak phases of the epidemic in 2020, worldwide dissemination and sharing.

The valve, consisting of two pieces, the Charlotte and the Dave valves, connects the mask to the oxygen dispenser allowing the patient to breath while guaranteeing the correct air insufflation pressure due to the tightness of the system.

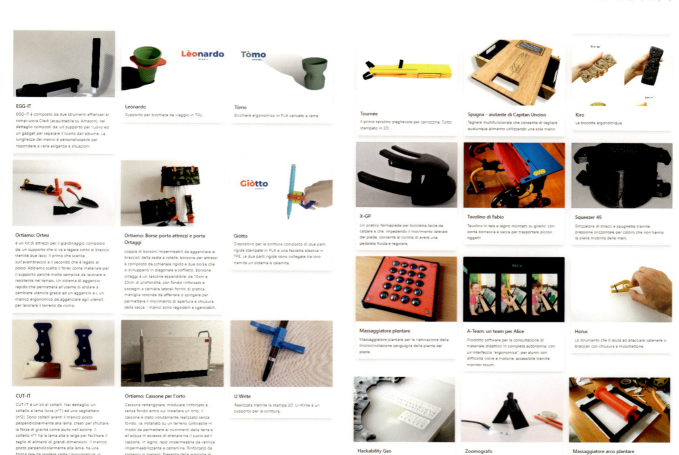

**EGG-IT**
EGG-IT è composto da due strumenti affiancati al rompi-uova Clack (acquistabile su Amazon), nel dettaglio composti da: un supporto per l'uovo ed un gadget per separare il tuorlo dall'albume. La lunghezza dei manici è personalizzabile per rispondere a varie esigenze e situazioni

**Lèonardo**
Supporto per bicchiere da viaggio in TPU

**Tòmo**
Bicchiere ergonomico in PLA caricato a rame.

**Tournée**
Il primo tavolino pieghevole per carrozzina. Tutto stampato in 3D.

**Spugna - aiutante di Capitan Uncino**
Tagliere multifunzionale che consente di tagliare qualunque alimento utilizzando una sola mano

**Kiro**
La biscotte ergonomique

**Ortiamo: Ortesi**
è un kit di attrezzi per il giardinaggio composto da: un supporto che si va a legare sotto al braccio tramite due lacci: il primo che scarica sull'avambraccio e il secondo che è legato al polso. Abbiamo scelto il foro come materiale per il supporto perché molto semplice da lavorare e resistente nel tempo. Un sistema di aggancio rapido che permetterà all'utente di andare a cambiare utensile grazie ad un aggancio a L un manico ergonomico da agganciare agli utensili per lavorare il terreno da vicino.

**Ortiamo: Borse porta attrezzi e porta Ortaggi**
coppia di borsoni impermeabili da agganciare ai braccioli della sedia a rotelle, borsone per attrezzi è composto da schienale rigido e due borse che si sviluppano in diagonale a soffietto. Borsone ortaggi è un tascone espandibile: da 10cm a 20cm di profondità, con fondo rinforzato e sostegni a cerniera laterali forniti di pratica maniglia rotonda da afferrare o spingere per permettere il movimento di apertura e chiusura della sacca, i manici sono regolabili e sganciabili.

**Giòtto**
Dispositivo per la scrittura composto di due parti rigide stampate in PLA e una fascetta elastica in TPE. Le due parti rigide sono collegate tra loro tramite un sistema a calamita.

**X-GP**
Un pratico fermapiede per bicicletta facile da calzare e che, impedendo il movimento laterale del piede, consente al ciclista di avere una pedalata fluida e regolare.

**Tavolino di Fabio**
Tavolino in tela e legno montato su girello, con porta borraccia e sacca per trasportare piccoli oggetti

**Squeezer 46**
Strizzacci di stracci e spugnette tramite pressione orizzontale per coloro che non hanno la piena mobilità delle mani.

**CUT-IT**
CUT-IT è un kit di coltelli. Nel dettaglio: un coltello a lama liscia (n°1) ed uno seghettato (n°2). Sono coltelli aventi il manico posto perpendicolarmente alla lama, creati per sfruttare la forza di gravità come aiuto nell'azione. Il coltello n°1 ha la lama alta e larga per facilitare il taglio di alimenti di grandi dimensioni. Il manico, posto perpendicolarmente alla lama, ha una forma tale da rendere salda l'impugnatura: in aggiunta è presente un elemento che va applicato a posteriori, nel punto che si desidera, per bloccare il movimento del pollice.

**Ortiamo: Cassone per l'orto**
Cassone rettangolare, modulare rinforzato e senza fondo entro cui installare un orto. Il cassone è stato volutamente realizzato senza fondo, va installato su un terreno coltivabile in modo da permettere ai nutrimenti della terra e all'acqua in eccesso di drenare tra il suolo ed il cassone. In legno, reso impermeabile da vernice impermeabilizzante e catramina. Rinforzato da sostegni in metallo. Presenta delle maniglie di sicurezza come appoggio aggiuntivo durante la lavorazione dell'orto.

**U Write**
Realizzata tramite la stampa 3D. U-Write è un supporto per la scrittura.

**Massaggiatore plantare**
Massaggiatore plantare per la riattivazione della microcircolazione sanguigna della pianta del piede.

**A-Team: un team per Alice**
Prodotto software per la consultazione di materiale didattico in completa autonomia, con un'interfaccia 'ergonomica', per alunni con difficoltà visive e motorie, accessibile tramite monitor touch.

**Horus**
Lo strumento che ti aiuta ad allacciare catenelle o braccialetti con chiusura a moschettone

**Hackability Geo**
La prima mappa 3D dell'Italia interattiva per bambini ciechi.

**Zoomografo**
Un economico ingranditore portatile per la lettura e la scrittura

**Massaggiatore arco plantare**
Riabilitazione muscoli per arco plantare

改造普通物品的开源项目。
Open source projects for adapting ordinary objects.

Hackability 组织
Hackability

## 2020 年荣誉提名奖
## Hackability 组织，2016 年
### 社会福利项目

Hackability 组织聚集了设计师、创客和数字工匠等能工巧匠，结合残疾人士的发明才能，针对这一群体的需求，超越现有产品的最初设计，将其改造为符合需求的全新产品，激发产品更多潜能。

通过共同设计、数字制造，使用 3D 打印机和开发板，公司、微观装配实验室和社区中心能够参与制作全新产品、改进现有产品或降低产品造价，由此促进社会凝聚与包容，培育新技能，打造更加包容的福利与文化服务。

## Honorable Mention 2020
## Hackability, 2016
### Social welfare project

Hackability brings together the skills of designers, makers and Digital craftsmen with the needs and inventiveness of people with disabilities, adapting existing products to need which were not contemplated in the original design, expanding the possibilities of the original objects.

Through co-design and Digital fabrication, the use of 3D printers and open-source boards, companies, fablabs and community centres can intervene in the realisation of products that do not yet exist, improve existing ones or make them cheaper, is a tool for developing social cohesion, inclusion, new skills, more inclusive welfare and cultural services.

# 住所无定式

## Limitless Domesticity

什么是室内？什么是室外？什么是客厅家具？"家"从哪里开始？在哪里结束？意大利设计不断地质疑各种观念的边界，催生了好多传奇性项目。汽车和床变成了插入式、灵活、可调整的人类住处，车床灯和手推车进入了家庭，设计师以各种可能的方式解构和重组着厨房。从 1972 年纽约举办的展览"意大利：新家庭景观"展示的模块化乌托邦概念，到大规模生产的长货架，意大利设计将所有可居住空间纳入到了"家"的定义中。

What is indoor and what is outdoor? What is an industrial tool and what is a piece of furniture for a living room? Where does "home" begin and where does it end? Italian design has long been questioning this kind of boundaries through projects that have become legendary. Cars and bedframes conceived as plug-in, flexible and editable human habitats, workshop lamps and carts entering domestic space, kitchens deconstructed and recomposed in all possible ways: from the modular utopian concepts shown in New York for the 1972 exhibition *Italy: A New Domestic Landscape* to mass-produced longsellers, products of Italian design have expanded the notion of home to all inhabitable spaces.

生活空间儿童床与设备的旧照。
Vintage photograph of Abitacolo with equipments.

Rexite 档案
Archivio Rexite

## ⋀ 1979 年 "生活空间" 儿童床

### 布鲁诺 · 穆纳里为 Robots 公司设计，1971 年

### 床

这不仅仅是一张床，而是一个系统，可以发挥想象力，安装生活必需的所有组件。

从字面上看，Abitacolo 是 "驾驶舱" 和 "居住地" 的双关。1971 年，穆纳里将 Abitacolo 具象化为一张床，上面可以安装书架、书桌、储物箱、衣架等不同的组件。

正因为结构非常简洁，"生活空间" 儿童床才如此功能强大、灵活，只有 51 公斤重，可轻松折叠放在运输箱中，为儿童在家中打造一片能够自由、趣味地想象和塑造的空间。

## ⋀ 1979 Abitacolo

### Bruno Munari, Robots, 1971

### Bed

More than a bed, a system where to install all necessary components of a life to be imagined.

Literally a pun with the ideas of "cockpit" and "place to inhabit", Abitacolo was conceived by Munari in 1971 as a bed structure on which different components can be installed, plugged in: bookshelves, a desk, storage boxes, a coat rack.

The extreme versatility and performativity of Abitacolo comes from the simplicity of its structure, only 51 kg heavy, easily collapsible in a shipping box, and aims to fulfil the need of kids to find a space of their own inside their family home, to be freely, playfully imagined and shaped.

生活空间儿童床作为音乐会舞台的旧照。
Vintage photograph of Abitacolo as a concert stage.

Rexite 档案馆
Archivio Rexite

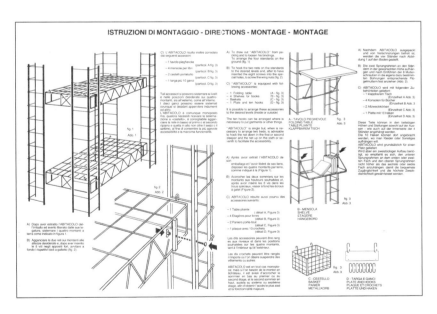

生活空间儿童床说明书。
Abitacolo instruction sheet.

ADI 设计博物馆档案
Archivio ADI Design Museum

生活空间儿童床结构的旧照，组装前折叠的样子。
Vintage photograph of Abitacolo structure, folded before assembly.

帕尔马大学传播学研究中心
CSAC, Università di Parma

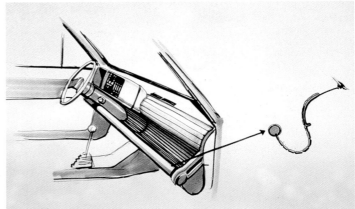

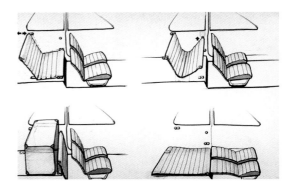

展示座椅套耐洗性的旧照。展示座椅套可拆卸性的旧照。手套箱和仪表板的旧照。
Vintage photographs demonstrating the washability of the seat coverings and the removability of the seat coverings, and showing the glove compartment and the dashboard.

ADI 设计博物馆档案
Archivio ADI Design Museum

座椅不同位置配置的仪表板区域的草图。
Sketches of different configurations in the positioning of the seats and of the dashboard area.

意大利设计档案馆
Archivio Italdesign

## ⋀ 1981 年 熊猫车

### 意大利设计 – 乔治亚罗公司为菲亚特公司设计 1980 年

### 车

在全球危机中诞生的标志性产品，熊猫车为各类用户提供了可居住的空间。

熊猫车作为全新的汽车概念，区别于当时已有的汽车类型，完全以内部的生活体验和功能为中心，摆脱了小型基本款汽车的所有传统。

座椅套可以拆卸，配有简单的锁定装置，可以针对不同用途做出调整。帆布收纳袋贯穿整个乘客舱前部。抛弃了传统的仪表板，在仪表盘外装上一个塑料外壳，而外壳本身是一个平面设计项目。

作为"强大的小型车"，熊猫车的出现为负载能力树立了新的标杆，但最重要的是，它既是生活方式的颠覆者，也是创造者。宣传强调其用户友好性（"……的朋友"）及受众的广度。

## ⋀ 1981 Panda

### Giorgetto Giugiaro – SIRP Italdesign [Italdesign Giugiaro], Fiat Auto, 1980

### Car

An instant icon born out of global crisis, Panda is a car conceived as an inhabitable space for all kinds of users.

Panda developed a totally new concept of car from existing mechanics, entirely centered on its interior living experience and versatility, getting rid of all traditional habits of the basic small car.

Seats covers are removable, a simple locking system allows them to offer a large number of different configurations for different uses, a canvas storage pocket runs the entire length of the front of passenger compartment, the conventional dashboard is deconstructed, with the instrumentation housed in a single plastic shell that is a graphic design project in itself.

Launched as "The great small car", since the beginning Panda set a new standard in load capacities, but most of all it was both the representation and the generator of a radical lifestyle change. Communication emphasised its user-friendly nature ("Friend of … "), and the width of its audience.

发布活动上的图片。
Launching campaign visuals.

菲亚特历史展览中心
Centro Storico Fiat

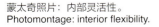

蒙太奇照片：内部灵活性。
Photomontage: interior flexibility.

菲亚特历史展览中心
Centro Storico Fiatt

广告照片
Advertising image

菲亚特历史展览中心
Centro Storico Fiat

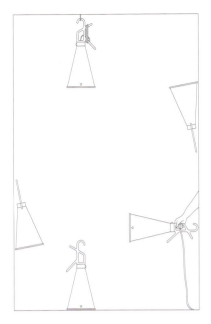

显示灯的各种可能位置的技术图纸。
Technical drawing showing different possible
positionings of the lamp.

Flos

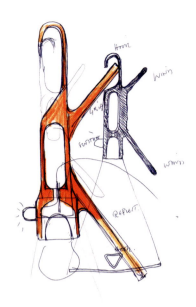

手柄的细节草图。
Detail sketch of the handle.

康斯坦丁·格西奇设计
Konstantin Grcic Design

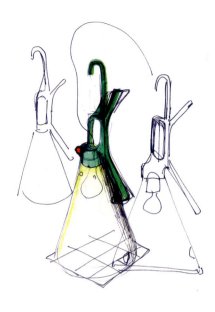

灯组件的解释草图。
Explanatory  sketch of the lamp components.

康斯坦丁·格西奇设计
Konstantin Grcic Design

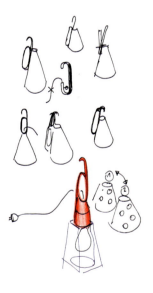

不同手柄设计方案的初步草图。
Preliminary sketch for different handle design
options.

康斯坦丁·格西奇设计
Konstantin Grcic Design

不同手柄设计方案的初步草图。
Preliminary sketch for different handle design
options.

康斯坦丁·格西奇设计
Konstantin Grcic Design

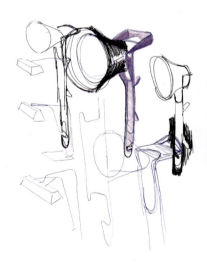

不同灯和手柄设计方案的初步草图。
Preliminary sketch for different lamp and
handle design options.

康斯坦丁·格西奇设计
Konstantin Grcic Design

## 2001 年"五月天" 照明设备

### 康斯坦丁 · 格西奇为 Flos 品牌设计，1998 年

### 灯具

受车间灯具的启发，"五月天" 照明设备打破了工作空间和生活空间之间的界限以及功能和美学之间的界限。

格西奇的塑料灯由一个挂钩 / 把手、一个开关、一根 5 米长的电线、一个半透明的截锥组成，截锥既是灯罩，又是支架，既实用又美观。灯可以手持、悬挂、放在桌面或地面。

"五月天"的形状和设计理念都很独特（名字与国际劳动节相呼应），既是操作性很强的工具，也是家用电器，赋予家庭空间特定的标志性身份。

## 2001 Mayday

### Konstantin Grcic, Flos, 1998

### Lamp

Inspired by workshop lamps, Mayday blurs the boundaries between workspace and living space, between functionality and aesthetics.

A plastic object made of a hook/handle, a switch, a 5-meter wire, a semi-transparent shortened cone at once a lampshade and a support, Grcic's lamp is as practical as it is aesthetic; it can be held, hung, table-top or floor-standing.

Mayday's shape and concept (and its name echoing International Workers' Day) make it both a fully operational working tool and a home appliance giving a specific and iconic identity to domestic spaces.

带轮咖啡桌
盖·奥伦蒂为 Fontana Arte 品牌设计，1980 年
Wheeled table
Gae Aulenti, Fontana Arte, 1980

ADI 设计博物馆档案
Archivio ADI Design Museum

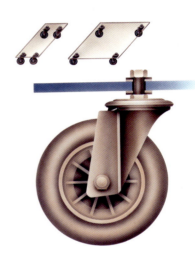

带有轴测图的细节图。
Detail drawing with axonometric views.

FontanaArte

## 1981 年荣誉提名奖 带轮咖啡桌

### 盖 · 奥伦蒂为 FontanaArte 品牌设计，1980 年

### 桌子

一辆工业推车成为家用室内设计的里程碑。

盖 · 奥伦蒂跳过技术图纸，凭直觉设计，然后便直接投入生产，呼应了 20 世纪初欧洲前卫艺术的现成创作过程。

将工厂运送玻璃板的推车转化为必备的咖啡桌，钢化玻璃板与黑色工业金属轮相结合，打破了生产空间和生活空间之间的界限。

## Honorable Mention 1981, Wheeled table

### Gae Aulenti, Fontana Arte, 1980

### Table

An industrial cart becomes a milestone in domestic interior design.

Going immediately from intuition to production without passing through technical drawing, the design by Gae Aulenti retraces the readymade creative process of European art avant-gardes of early 20th century.

Translating the wheeled platforms carting glass slabs around the factories into an essential coffee table, a tempered glass slab is combined to black-painted industrial metal wheels, breaking the boundaries between production and living space.

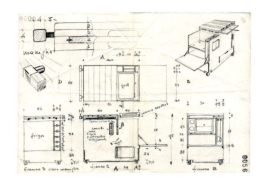

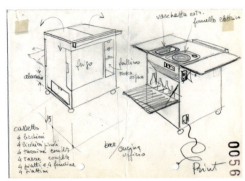

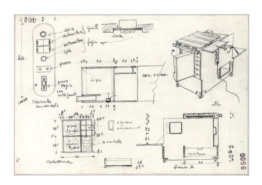

乔·科隆博的细节草图。
Detail sketches by Joe Colombo.

Boffi 提供
Courtesy Boffi

设备齐全的迷你厨房的旧照。
Vintage photograph of fully equipped Minikitchen.

Boffi 提供
Courtesy Boffi

## 2022 年终身成就奖，迷你厨房
### 乔 · 科隆博为 Boffi 品牌设计，1963 年
### 厨房模块

将厨房缩小成一个立方体，可以随处放置，这是对意式设计智慧的颂歌。

迷你厨房将必备的烹饪、辅助和存储功能集中在单一的物体上，变成一个方便移动的"手推车"，可以放置在 1960 年意大利和欧洲的现代公寓中。在 1972 年纽约现代艺术博物馆举办的展览"意大利：新家庭景观"中展出。

迷你厨房最初配备的是不锈钢燃气灶、胶合板和白色涂装。自 2014 年以来，改成船用胶合板生产，并带有电磁炉，可以放到室外使用。

## Career Award 2022 Minikitchen
### Joe Colombo，Boffi，1963
### Kitchen module

An entire kitchen shrunk into one single cube, to be placed anywhere, an ode to the smartness of Italian design.

Minikitchen aimed to condensate all functions of cooking, support and storage required to kitchens in one single object, a wheeled "cart", easy to move and place within modern apartments of 1960 Italy and Europe. It was part of *Italy: A New Domestic Landscape* exhibition (1972) at MoMA in New York.

Originally released in plywood with stainless steel gas stove, Mnikitchen was then given a white livery and since 2014 it is produced in marine plywood – with induction plates – so as to extend its realm also to the outside of houses.

上：多板配置的数字草图。
下：单板配置和双板配置的数字草图。
Top: digital sketches of different configurations.
Bottom: digital sketches of single-plate and
double-plate configuration.

阿德里亚诺设计提供
Courtesy Adriano Design

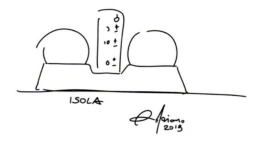

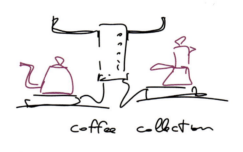

## 2022 年"秩序"电磁炉

### 阿德里亚诺设计公司为 Fabita 公司设计，2019 年

### 电磁炉

这款烹饪用具在 21 世纪 20 年代属于颠覆性设计。

同样是烹饪用具，但在迷你厨房问世 60 年后才诞生：这一次电磁炉几乎是独立式的，可以放在不同距离和各种表面上。

电磁炉也可以单独使用，关闭后，可以存储在可见的位置，为墙壁和桌子增添景观。

## 2022 Ordine

### Adriano Design, Fabita, 2019

### Induction hobs

Radical cooking space design for the 2020s.

A cooking system again, but 60 years after Minikitchen: this time the (induction) plates are almost freestanding objects that you can place at different distances and on different surfaces.

The plates can also work singularly, and once turned off, after having operated as a functional archipelago, they can be stored but left visible, creating a landscape for walls and tables.

# 发明与创造

Invention

意大利从不缺乏设计师兼发明家。这些人凭着一股钻研到底的精神，发现新材料，观察市场、技术和社会现象。例如思考新光源的出现、研究 18 世纪某个光学解决方案、探索新的餐饮仪式等等。这些钻研活动常常从根本上改变了设计的构思过程和理解方法。作为发明家，他们将这些活动变成了革命性的时运，在工业生产者的支持下，为集体文化和生活方式的创新做出了杰出贡献。

A spirit of ceaseless research dominates the activity of designer-inventors, figures densely populating the Italian scenario: the discovery of new materials, or the observation of phenomena – be them market, technique or behaviour-driven, like the appearance of a new light source, the study of an 18th century optical solution or the detection of a new eating ritual – often triggered radical shifts in the way design processes were conceived and understood. The result of the inventor's work is what turned these events into revolutionary moments, supported by industry and producers making possible an outstanding contribution to innovation of collective culture and lifestyle.

带原包装的"仔仔"旧照。
Vintage photograph of Zizì with original packaging.

ADI 设计博物馆档案
Archivio ADI Design Museum

造型用的原始测试模型。
Raw testing model for molding.

布鲁诺·穆纳里收藏品 — CLAC Cantù/ADI 设计博物馆档案
Collezione Bruno Munari - CLAC Cantù/Archivio ADI Design Museum

全新泡沫塑胶玩具，*Domus* 杂志，第 277 期， 1952 年 12 月。
A new foam rubber toy. *Domus* n.277, December 1952.

*Domus* 杂志档案 © Editoriale *Domus* 联合股份有限公司
Archivio *Domus* © Editoriale *Domus* Spa

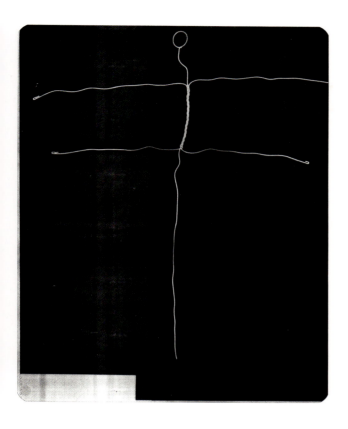

"仔仔"铜线结构的 X 射线图。
X-ray of Zizì copper wire structure.

私人档案
Archivio Privato

"仔仔"专利图
Patent drawing of Zizì.

意大利国家中央档案馆
Archivio Centrale dello Stato

## 1954 年"仔仔"玩具

### 穆纳里为 Pigomma 公司设计，1953 年

### 强化泡沫橡胶玩具

用发泡橡胶和铜线打造的玩具兼工具，能够激发儿童的想象力。

灵活的铜线骨架，泡沫橡胶身子，一只聪明的猴子便诞生了，可以变换不同的姿势，手能够抓住东西。

仔仔由倍耐力旗下的 Pigomma 开发，与孪生兄弟 Meo 猫（Gatto Meo）一起，构成一系列玩具组合，能够激发儿童的想象力。仔仔是一款开放的工具，是开启儿童想象力的平台。这种设计态度就能够让画家巴勃罗 · 毕加索称穆纳里为"哲学家"。

## 1954 Zizì

### Bruno Munari, Pigomma, 1953

### Reinforced foam rubber toy

Foam rubber and copper wire to generate both a toy and a tool for childrens' imagination.

On a flexible copper wire structure, a foam rubber body is molded in the shape of a witty monkey that can take different positions, its hand capable of closing and grabbing things.

Generating the imaginary of an entire collection of toys by Pirelli – branded Pigomma – together with its twin Gatto Meo (Meo the Cat), Zizì is a toy that acts as an open tool, a starting platform for the imagination of children. For such attitude to design, painter Pablo Picasso would call Munari "a philosopher".

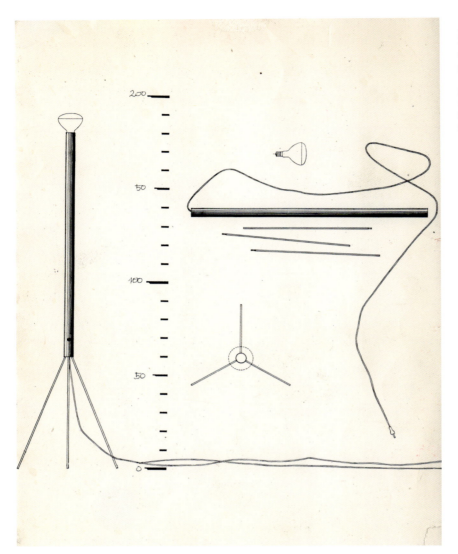

"发光者"灯具，彼得罗·基耶萨设计，
Fontana Arte 目录，20 世纪 40 年代。
Luminator by Pietro Chiesa,
Fontana Arte catalogue, 1940s.

Fontana Arte 品牌
Fontana Arte

Tubino，阿希尔·卡斯蒂廖尼和皮耶尔·贾科莫·卡斯蒂廖尼，1949 年。
Tubino, Achille and Pier Giacomo Castiglioni, 1949.

阿希尔·卡斯蒂廖尼基金
Fondazione Achille Castiglioni

Bulbo，阿希尔·卡斯蒂廖尼和皮耶尔·贾科莫·卡斯蒂廖尼，1957 年。
Bulbo, Achille and Pier Giacomo Castiglioni, 1957.

阿希尔·卡斯蒂廖尼基金
Fondazione Achille Castiglioni

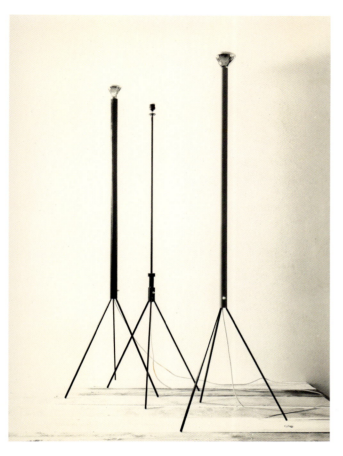

"发光者"灯具，20 世纪 50 年代。
Luminator, 1950's.

阿希尔·卡斯蒂廖尼基金
Fondazione Achille Castiglioni

"Dive Luminose"展览图片，
Centrodomus，米兰，1981 年。

Image of the Dive Luminose
exhibition, Centrodomus, Milan,
1981.

## 1955 年 "发光者"灯具

### 阿希尔·卡斯蒂廖尼和皮耶尔·贾科莫·卡斯蒂廖尼为 Gilardi & Barzaghi – Flos 设计，1955 年

**带间接照明的落地灯**

光秃秃的电灯泡本是经典耐用的工业产品，搭配轻巧而稳定的框架，便进入了寻常百姓家。

"可以说，'发光者'是为了 1955 年金圆规奖而生，或者更确切地说，响应了意大利工业对'实用形状'的需求。早一年文艺复兴百货（Rinascente）宣布组织这个著名奖项后，便催生了这个需求。"

灯泡的内部镜面反射器在传递光束的同时防止眩光，在当时这种现象非常新奇，也是"发光者"的基本特征原理，符合现代家庭的普遍要求：将形式与功能结合起来。

## 1955 Luminator

### Achille and Pier Giacomo Castiglioni
### Gilardi & Barzaghi - Flos, 1955

**Floor lamp with indirect light**

A lightweight yet stable framework introduces an exposed light bulb – a typically industrial and long-lasting product – into the world of domestic light sources.

"It can be said that the Luminator was created specifically for the Compasso d'Oro of 1955, or rather it arose as our response to the demand for a 'useful shape' for the Italian industry prompted by the famous award announced by Rinascente, just the year before."

The bulb's internal mirror reflector which conveyed the light beam and protected against glare, a novelty at the time, was the basis of the essential character of the Luminator in accordance with the principle, typical of the modern home, that combined form with function.

1960 年第六届金圆规奖目录中的旧照。
Vintage photograph from the catalogue of the 6th Compasso d'Oro 1960.

ADI 设计博物馆档案
Archivio ADI Design Museum

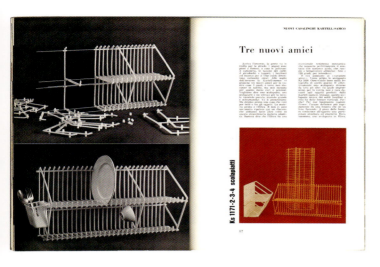

Qualità 杂志，第 11 期，1960
年冬，Kartell-Samco。
Qualità, n.11, winter 1960,
Kartell-Samco.

Kartell 博物馆
Museo Kartell

L'Europeo 杂志，第 46 期，1961
年 11 月，Rizzoli。
L'Europeo, n. 46, November
1961, Rizzoli.

私人收藏
Private collection

## 1960 年 KS 1171/2 型滤碗器

### 科隆比尼（Colombini）为 Kartell 公司设计，1960 年

### 可移动滤碗器

这款产品设计低调但堪称典范，主体与安装配件的完美结合，颠覆了
传统意义的厨具。使用当时的前卫材料低压聚乙烯，充分实现了尺寸
的灵活性，使性价比最大化。

基于这个体系，滤碗器的长度可以灵活调整，适应二战后时期小至迷
你厨房、大至中产家庭的各种空间。

## 1960 KS 1171/2

### Gino Colombini, Kartell-Samco, 1960

### Removable dish drainer

A modest but exemplary lesson in design, it renewed the
traditional kitchen tool thanks to the close relationship between
the module and the mounted units. This resulted in maximum
flexibility of use, both in terms of size and cost-effectiveness, by
utilizing low-pressure polyethylene, a highly innovative material
at the time.

The system made it possible to have the same dish drainer in
different lengths making the product more compatible with the
varying domestic spaces of the post-war period, ranging from
the middle-class home to the minimal kitchen.

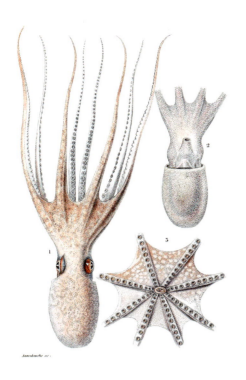

麝香章鱼。
Musky Octopus.

92136286 © cc0images | Dreamstime.com

《20 件充满诗意反应的物品》，朱利奥·亚凯蒂和马泰奥·拉尼设计，保拉·乔里编辑，Editrice Compositori 出版社，2002 年。
*20 oggetti a reazione poetica*, Giulio Iacchetti e Matteo Ragni designer, edited by Paola Jori, Editrice Compositori 2002.

朱利奥·亚凯蒂档案
Archivio Giulio Iacchetti

矩阵模具。
Matrix mould.

潘多拉设计（捐赠）/ADI 设计博物馆档案
Pandora design (donor) / Archivio ADI Design Museum

木制原型和最终产品。
Wooden prototypes and final products.

朱利奥·亚凯蒂档案
Archivio Giulio Iacchetti

朱利奥·亚凯蒂和马泰奥·拉尼与"小章鱼"一次性叉勺的合照。
Portrait of Giulio Iacchetti and Matteo Ragni with Moscardino.

朱利奥·亚凯蒂档案
Archivio Giulio Iacchetti

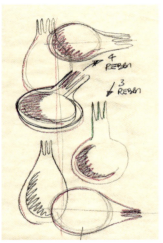
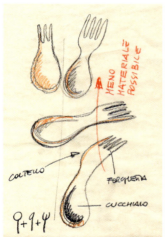
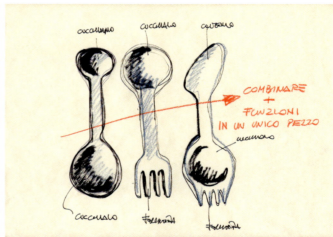

草图。
Sketches.

朱利奥·亚凯蒂档案
Archivio Giulio Iacchetti

## ⋀ 2001 年，"小章鱼"一次性叉勺

**朱利奥·亚凯蒂和马泰奥·拉尼为**
**Pandora Design 公司设计，2000 年**

**多功能餐具套装**

将叉子和勺子的功能融于一身，使得用餐成为了休闲而欢乐的时刻。

意大利语 Moscardino 是一种小章鱼的名字，形状与这个设计相呼应，既神秘又实用。

这套餐具最初是为开胃酒仪式而设计，使用可降解的生物塑料制成，后来改用不锈钢，证明了设计的质量之高，适用于日常生活。

## ⋀ 2001 Moscardino

**Giulio Iacchetti, Matteo Ragni**
**Pandora Design, 2000**

**Multi-purpose cutlery set**

It combines in a single object the function of a fork and a spoon, transforming the act of eating into an informal and convivial moment.

"Moscardino" in Italian is the name of a small octopus whose shape is echoed in this object, at once enigmatic and functional.

Conceived for the ritual of the aperitif and initially conceived in mater-bi, a compostable bioplastic, it was later produced in steel, testifying to the design quality that extends its function to everyday life.

 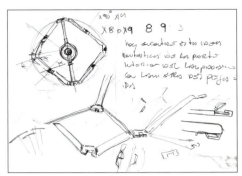

"希望"吊灯内部结构草图。
Sketches of the internal structure of Hope.

弗朗西斯科·戈麦斯·帕斯
Francsico Gomez Paz

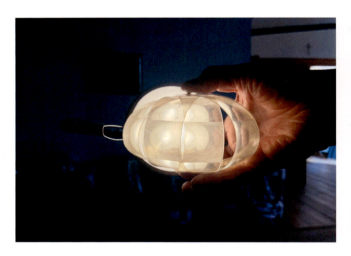  

## ⋀ 2011 年 "希望"吊灯

弗朗西斯科 · 戈麦斯 · 帕斯和保罗 · 里扎托为
Luceplan 公司设计, 2009 年

### 灯具

设计者在探索新的吊灯类型时, 设计了"希望"吊灯, 一个集装饰性、象征性和功能性于一身的典范。这款设计不再拘泥于模仿, 通过了光学实验, 从科学和艺术研究中获得灵感。

这款轻质吊灯由工业常规的模块化单元组成, 形状是透明的叶子, 增加了中央光源, 能够照亮整个空间。

经过多次试验, 将菲涅耳透镜原理应用到薄薄的雕刻叶子上, 增加光强度, 使照射范围最大化。

## ⋀ 2011 Hope

Francisco Gomez Paz, Paolo Rizzato
Luceplan, 2009

### Lamp

Hope is the result of the interest in renewing the chandelier typology – a paradigmatic example of an object of ornamental, symbolic and functional value – through optical experimentation taking its cue from scientific and artistic research, overcoming a mimetic attitude.

A luminous and lightweight volume consisting of a modular unit which is repeated industrially, forming the transparent leaves that multiply the central light source, fills the environment.

After many trials the principle of the Fresnel lens was translated onto thin sculpted leaves which increase light intensity and direct it as far as possible.

海洋硅藻的光学显微镜观察情况和
透镜的制作。
Light microscopy of marine
diatoms and the making of the
lense.

Editoriale Lotus 提供
Courtesy Editoriale Lotus

"希望"吊灯的形成、早期探索和
最终方案。
The making of Hope, previous
attempts and final solution.

Francisco Gomez Paz 和
Editoriale Lotus 提供
Courtesy Francisco Gomez
Paz and Editoriale Lotus

# 家居生活

## Families

意大利设计常常从组合角度展开构思，让用户选择不同元素组合成定制化作品，或者呈现由复杂的制程产生的不同结果。这些设计特点鲜明，放到不同的组合中，非但没有失去特征，反而变得更加突出，说明其可塑性很强，能够适应不同语境。更有甚者创作过程摆脱传统束缚，不再由设计师完全掌控，而是向公众开放：结果不再是落锤定音的成品，而是拥有一系列可能性，让用户参与最终的决断。

"家居生活"往往也有一个真实的叙事角色：构成元素仿佛就是场景中的人物，一如托洛梅奥系列灯具或里卡尔多 · 达利西研究的那不勒斯咖啡机。

In addition to stand-alone objects, Italian design has often reasoned in terms of 'Families', different elements that can be selected or combined by users in customized compositions, or that express different possible outputs of complex research paths. These are projects with a strong identity that are declined in a series of variations without losing their character, on the contrary coming out stronger, because they prove to be elastic and adaptable to the context. Moreover, the creative process, which tradition wants to be exclusively in the hands of the designer, is here shared with the public: no longer a closed, finished project, but a range of possibilities that involve the user, making him participate in the final choices.

The 'Families' also often have a real narrative character: the elements that constitute them appear to us like characters in a scene, as in the case of the Tolomeo Lamp or Riccardo Dalisi's Research on Neapolitan Coffee Pots.

"条纹家族"扶手椅草图，奇尼·博埃里，20 世纪 70 年代 。
Sketches of the Strips family. Cini Boeri, 1970s.

Seven Salotti 股份有限公司 - Arflex
Seven Salotti Spa - Arflex

"条纹家族"扶手椅草图，奇尼·博埃里，20 世纪 70 年代。
Sketches of the Strips family. Cini Boeri, 1970s.

Seven Salotti 股份有限公司 - Arflex
Seven Salotti Spa - Arflex

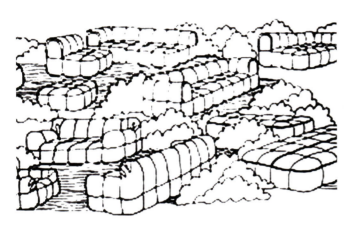

## ⋀ 1979 年 "条纹家族" 扶手椅

### 奇尼 · 博埃里和劳拉 · 格里齐奥蒂为 Arflex 品牌设计
### 1972 年

### 软垫家具系列

奇尼 · 博埃里在尝试包装沙发时开启了这个项目，结合实用的罩子和模块化系统，打造了一个家族，包含扶手椅、沙发、床以及沙发床。

罩子如同真实的外皮，可以脱除、清洗、更换和整修，带有拉链，可以像衣服一样穿在聚氨酯的本体上。

床和沙发床都成为了抽象的元素，用于打造新型的多功能空间：打开睡袋的拉链，折叠靠背，环境便会发生变化。

"条纹家族"今天仍在生产，多年来不断发展演变，呼应了奇尼 · 博埃里所期望的工业节奏：每年重新审视产品，而非急于更新。

## ⋀ 1979 Strips

### Cini Boeri with Laura Grizziotti
### Arflex, 1972

### Series of upholstered furniture

The formal exercise underlying the project – which started with Cini Boeri trying to wrap up a sofa – is combined with the practicality of the cover and the modularity of the system, creating an open system, a family of different characters, that offers the possibility of integrating armchairs, sofas, beds and sofa beds.

The cover is used as a skin, to be stripped off, washed, changed and refitted, it closes with a zip fastener, just like a dress, on the polyurethane body.

Beds and sofa beds become elements with an abstract image that create a new type of multifunctional space: an environment that changes when opening the zipper of the sleeping bag and folding the backrest down.

Still in production today, the Strips family has expanded and changed over the years, an evolution which responds to an industrial rhythm desired by Cini Boeri, reviewing the product instead of rushing to replace it with new products every year.

"条纹家族"扶手椅的广告活动照片。摄影：阿尔多·巴洛，艺术总监吉安卡洛·伊利普兰迪。条纹家族扶手椅，Arflex，20 世纪 70 年代。
Photographs of the advertising campaign for the Strips family. Photo by Aldo Ballo, Art Director Giancarlo Iliprandi. La famiglia degli Strips, Arflex, 1970s.

Seven Salotti 股份有限公司 - Arflex
Seven Salotti Spa - Arflex

绵羊椅，"条纹家族"扶手椅，Arflex，20 世纪 70 年代。
Pecorelle, La famiglia degli Strips, 1970s.

阿弗莱斯历史档案馆
Archivio Storico Arflex

咖啡机原型。
Prototypes of coffee pots.

ADI 设计博物馆档案
Archivio ADI Design Museum

## ⋀ 1981 年 那不勒斯咖啡机

### 里卡尔多 · 达利西为 Alessi 品牌进行的一项研究，1979—1981 年

### 咖啡机

达利西说，那不勒斯咖啡机有一个特点：要像杂技演员，倒过来使用。咖啡机变成了一个角色，咖啡冲泡仪式变成了一种叙述，从那不勒斯传统中著名的普尔奇内拉面具缓缓说起。

各类原型是相同主题的不同变体：基于角色的多样性，综合提炼出适合市场的标准化物品。

最终产品沿用了那不勒斯咖啡机的经典形状，并结合当代美学线条和工业生产需求进行升级。

## ⋀ 1981 Research for the production of a Neapolitan coffee pot

### Riccardo Dalisi, Alessi, 1979-81

### Coffee pot

The Neapolitan coffee pot has a peculiarity, explains Dalisi: when you use it you have to turn it upside down, as if it were an acrobat. His coffee pot then becomes a character and the coffee ritual becomes a narrative, starting with Pulcinella, the famous mask of the Neapolitan tradition.

The prototypes are all variations on the theme: from this multiplicity of characters, we arrive at the synthesis, represented by the standard object, designed for the market.

In the final product, the classic shape of the Neapolitan coffee pot is respected and at the same time updated to adhere with the contemporary aesthetics and the needs of industrial production.

咖啡机原型。
Prototypes of coffee pots.

ADI 设计博物馆档案
Archivio ADI Design Museum

草图。
Sketch.

里卡尔多·达利西档案
Riccardo Dalisi Archive

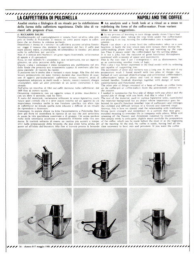

里卡尔多·达利西，普尔奇内拉咖啡机，*Domus* 杂志，第 617 期，1981 年 5 月。
Riccardo Dalisi, La caffettiera di Pulcinella, in *Domus*, n.617, May 1981.

*Domus* 杂志档案 ©Editoriale *Domus* 联合股份有限公司
Archivio *Domus* © Editoriale *Domus* Spa

1981 年 那不勒斯咖啡机，
里卡尔多·达利西为 Alessi 品牌进行的一项研究，
1979—1981 年。
咖啡机。
1981 Research for the production of a
Neapolitan coffee pot,
Riccardo Dalisi, Alessi, 1979-81,
Coffee pot.

ADI 设计博物馆档案
Archivio ADI Design Museum

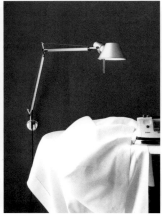

不同环境中不同款式的托洛梅奥系列灯具，20 世纪 80 年代末：用于案头工作、阅读、学习、缝纫。
Tolomeo in different versions and settings, late 1980s: Digital and analogical work, reading, studying, sewing.

ADI 设计博物馆档案
Archivio ADI Design Museum

## ⋀ 1989 年 托洛梅奥系列灯具

### 米凯莱 · 德卢基和吉安卡洛 · 法西纳
### 为 Artemide 品牌设计，1987 年

### 灯具系列

托洛梅奥系列灯具是一盏灯，更是一道公式、一种产品哲学。设计非常灵活，适用于各种场景，每个组件都可以变成独立的一盏灯。

托洛梅奥系列灯具可以创设不同环境，也能适应各种环境，能够引导光线，将光指向任何方向。

经典造型在不同角色中逐渐消融，不再拘泥于同一个模型，形成了一个产品系列。在工作模式与生活方式出现混合形态的时代，这是一种基本的价值观。

## ⋀ 1989 Tolomeo

### Michele de Lucchi and Giancarlo Fassina
### Artemide, 1987

### Series of lamps

Rather than a lamp Tolomeo is a formula, a philosophy of the product. It has a very pliable personality and can be used anywhere, and each of its components can be turned into an independent lamp.

Tolomeo gives the possibility of directing the light and moving the source in any direction thanks to a mechanism specifically intended to create and adapt to dynamic environments.

Its iconic shape declines in many different characters, exceeding the variations of a model, to create a family of products, a fundamental value at a time when new hybrid forms of work and lifestyle had already been established.

艾略特·厄威特，托洛梅奥，2001 年。
Elliott Erwitt, Tolomeo, 2001

Artemide 品牌
Artemide

托洛梅奥微型台灯的
原型和最终设计。
Prototype and final
version of Tolomeo
Micro.

摄影：乔凡娜·拉拉塔
/ 米凯莱·德卢基档案
Photo by Giovanna
Lalatta / Archivio
Michele de Lucchi

艾略特·厄威特，《向
光致敬》：米凯莱·德
卢基，2015 年。
Elliott Erwitt, A Tribute
to Light: Michele De
Lucchi, 2015

Artemide 品牌
Artemide

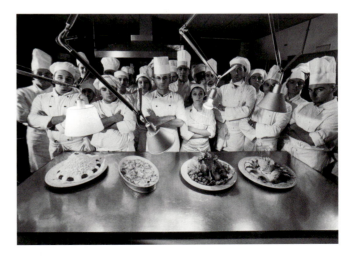

艾略特·厄威特，《向光致敬》：米凯莱·德卢基，2015 年。
Elliott Erwitt, A Tribute to Light: Michele De Lucchi, 2015.

Artemide 品牌
Artemide

△ 2020 年 "安排" 照明设备
迈克尔 · 阿纳斯塔夏季斯为 Flos 品牌设计，
2018 年

照明设备

△ 2020 Arrangements
Michael Anastassiades,
Flos, 2018

Lighting devices

"安排"照明设备组件的各种组合。
Possible combinations of the
Arrangements elements.

ADI 设计博物馆
ADI Design Museum

不同组合和不同环境中的"安排"照明设备。
Arrangements in different combinations and settings.

摄影：桑蒂·卡莱卡 / FLOS 档案
Photograph by Santi Caleca / Archivio FLOS

## ⋀ 2020 年 "安排" 照明设备

### 迈克尔 · 阿纳斯塔夏季斯为 Flos 品牌设计，2018 年

### 照明设备

线性的 LED 元件极致简约，让无数种可能的组合都具有独特性。

线条、折线、圆圈、水滴和方形等五种基本形状的 LED 灯，尺寸不一，轻松组合在一起，便能打造出无限种可能，既有最简单的灯具，也有窗帘灯和吊灯。

"这是介于发光雕塑和工业设计之间的作品"，或者正如阿纳斯塔夏季斯所说，是介于设计和珠宝之间的作品。不同形状之间达到相互平衡，仅仅是简单的"咔嗒一声关闭"和"咔嗒一声打开"，"安排"就迅速定义了照明设计的新原型。

## ⋀ 2020 Arrangements

### Michael Anastassiades, Flos, 2018

### Lighting devices

The extreme simplicity of linear LED elements guarantees the uniqueness of countless possible compositions.

Five fundamental LED shapes – line, broken line, circle, drop and square – in different sizes can be easily joined in a potentially infinite catalogue of combinations, from the simplest lamp to curtains or chandeliers.

"Midway between a light sculpture and an industrial design object" – or between design and jewellery, as Anastassiades says – through an easy "click to close" and "click to join" system based on the balancing of each shape on another, Arrangements soon defined a new archetype in lighting design.

转折点

Turning Point

一些项目在设计史上成为转折点，标志着其问世前后的迥然不同，这些标志性设计的作用往往是金圆规奖关注的焦点。真正的游戏规则改变者，一经问世，就重新定义了人们对某类物品的认识（如折叠手机的问世，如便携式电视机不再是笨重的家具，如豆袋变身座椅，反之亦然），它们同时重新定义了人们的生活习惯、生活方式。甚至有些时候，它们几乎催生了全新品类，或者至少塑造了标识性形象。

Some projects have worked as 'Turning Points' in the history of design, marking a before and after, and their role has often been remarked by the Compasso d'Oro. True game changers, that upon their appearance redefined the way we used to think about certain typologies of objects (by introducing the flip mechanism in the telephone, by making the television a portable object and no longer a piece of furniture, by turning a bean bag into a chair, and vice versa), or redefined ways of performing habitual actions. At other times, they have almost spawned new typologies, or at least made their image iconic.

电视机旋钮的俯视图。
View of the knobs from above.

Brionvega 档案
Archivio Brionvega

1963 年米兰贸易展览中心 Brionvega 展台的图片。
Images of the Brionvega stand at the Milan Trade Fair, 1963.

由米兰展览基金会历史档案提供
Courtesy of Archivio Storico Fondazione Fiera Milano

Algol 电视机，马尔科 · 扎努索和理查德 · 萨珀为 Brionvega 品牌设计，1964 年。
Algol television set designed by Marco Zanuso with Richard Sapper for Brionvega in 1964.

Brionvega 档案
Archivio Brionvega

## ⋀ 1962 年 "多尼" 便携式电视

### 马尔科 · 扎努索和理查德 · 萨珀为 Brionvega 品牌设计，1962 年

### 电视机

"多尼"为现代家庭注入了新活力，带来了全新的审美享受：它便于携带，替代了电子管电视机。

其组件安装在晶体管周围，并分为四组，逐件组装，便于拆卸和检查维修。

扎努索和萨珀为意大利家电公司 Brionvega 设计了多款型号后，最终确定了该设备的便携性。带上你的电视机，超级互联就在眼前。

## ⋀ 1962 Doney

### Marco Zanuso, Richard Sapper, Brionvega, 1962

### Television set

Doney introduced a new dynamic use and a new aesthetic to the modern domestic landscape: it allowed portability and represented an alternative to the imposing presence of the valve television set.

The components were mounted in the available space around the tube and separated into four groups assembled by pieces, facilitating disassembly and inspection for repairs.

The research developed with subsequent models designed by Zanuso and Sapper for Brionvega led to the device's definitive mobility. Take your tv set with you, hyperconnectivity is around the corner.

组装前部件的分解图。
Exploded view of the constituent parts
before assembly.

Brionvega 档案
Archivio Brionvega

标明组件的垂直剖面技术图纸。
Technical drawing of the vertical section with
indication of the components.

巴莱尔纳当代档案馆，马尔科·扎努索基金
Balerna, Archivio del Moderno, Fondo Marco
Zanuso.

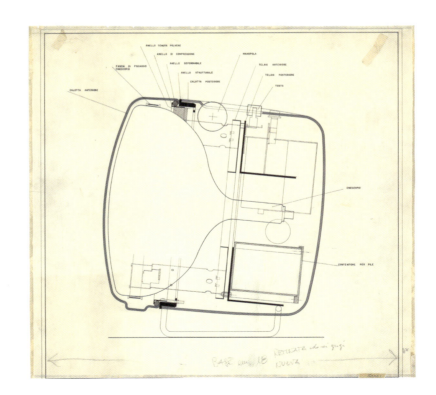

掌上的"蟋蟀"电话广告活动，1967 年。
Grillo the telephone in the palm of your hand
advertising campaign, 1967.

*Domus* 杂志档案 © Editoriale *Domus*
联合股份有限公司
Archivio *Domus* © Editoriale *Domus* Spa

## ⋀ 1967 年 "蟋蟀" 电话

### 马尔科 · 扎努索和理查德 · 萨珀为西门子公司设计，1966 年

### 电话

"蟋蟀"电话采用人体工学设计，提供全新、精简的使用方式，能够与用户一对一互动，乐趣十足，催生了新的行为方式。

在当前手机的非物质中立性中，有些东西注定要消融，"蟋蟀"电话保留了最后的痕迹，是电子元件小型化过程中的一种标志性中间步骤。

设计重点关注占用空间，与传统电话相比，"蟋蟀"电话非常小巧；主机和麦克风之间铰接接头方案也是设计重点。

## ⋀ 1967 Grillo

### Marco Zanuso with Richard Sapper Siemens, 1966

### Telephone

The attention paid to the ergonomics and a new, more streamlined method of use, blended with the playful, one to one interaction with the user which produced new forms of behaviour.

Grillo has revealed itself as the last trace of something destined to disappear in the immaterial neutrality of current mobile phones, representing a sort of intermediate step in the process of miniaturization of electronic components.

The design emphasis focused on the space taken up, which was very small compared to traditional telephones, and on the solution of the hinged joint between the main body of the device and the microphone.

掌上的"蟋蟀"电话广告活动，1968 年。
Grillo the telephone in the palm of your hand advertising campaign, 1968.

*Domus* 杂志档案
© Editoriale *Domus* 联合股份有限公司
Archivio *Domus* © Editoriale *Domus* Spa

马尔科·扎努索和理查德·萨珀一起研究"蟋蟀"电话，1966 年。
Marco Zanuso and Richard Sapper at work on the Grillo telephone, 1966.

摄影：毛罗·马塞拉
威尼斯建筑大学—项目档案，毛罗·马塞拉基金。
Photo by Mauro Masera. Università IUAV di Venezia — Archivio Porgetti, Fondo Mauro Masera

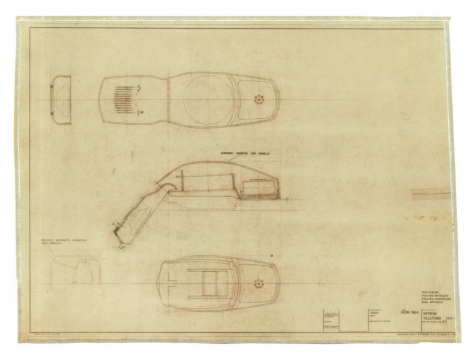

技术图纸、立面图和剖面图。
Technical drawings, elevations and sections.

巴莱尔纳当代档案馆，马尔科·扎努索基金
Balerna, Archivio del Moderno, Fondo Marco Zanuso.

旧照和广告
Vintage photographs and advertisements

Zanotta 提供
Courtesy Zanotta

## ⋀ 2020 年终身成就奖 豆袋沙发

### 皮耶罗 · 加蒂、切萨雷 · 保利尼、佛朗哥 · 泰奥多罗为 Zanotta 品牌设计，1968 年

### 座椅

"请坐。""这里只有一个豆袋。""是的，没错。"

意大利设计师三人组想到了一个简单而颠覆性的操作：宣布豆袋就是一把座椅。这个设计充分显示了 20 世纪 60 年代流行艺术的精神，将人造产品、漫画和商业语言玩弄于股掌之间，解构了传统的形状和类型。

加蒂、保利尼和泰奥多罗在袋子里装满聚苯乙烯球，变成一把椅子，袋子充当椅套。这是个极具功能性的概念：聚苯乙烯球的自然重力特性确保椅子的稳定性和灵活舒适性，形状是一种近乎抽象的色团，正如不同广告中所展示的那样，适用于任何环境。

## ⋀ Career Award 2020, Sacco

### Piero Gatti, Cesare Paolini, Franco Teodoro Zanotta, 1968

### Seat

"Please take your chair." "I can only see a bean bag." "Yes. Exactly."

Fully participating in the spirit of 1960s Pop Art, playfully valorizing artificial products, comics and commercial language, and deconstructing traditional shapes and typologies, the Italian designer trio would imagine a simple yet radical operation: declaring that a bean-bag is a chair.

Born from filling a bag – becoming an upholstery – with polystyrene balls, *Sacco* by Gatti, Paolini, Teodoro is an extremely versatile concept: stability and adaptive comfort are granted by the natural gravity-based disposition of polystyrene balls, and the shape becomes an almost abstract color volume, ready to fit any situation as shown in different advertisements.

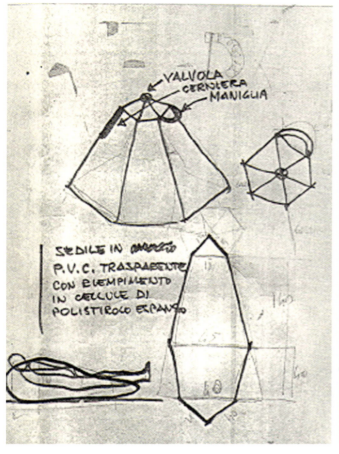
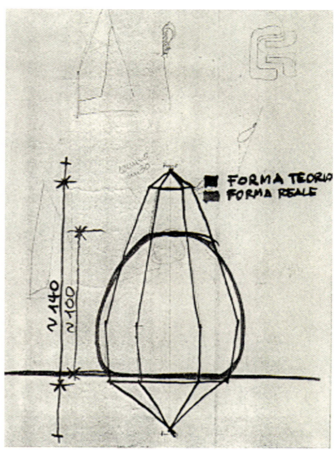

项目原始草图。透明 PVC 豆袋椅的初步配置草案，其中填充聚苯乙烯球。
Original project sketches. Preliminary configuration drafts for bean-bag chair in transparent PVC to be filled with polystyrene balls.

Zanotta 提供
Courtesy Zanotta

马尔科·卡内帕绘制的数字草图，展示了 Vespa 设计从早期车型到 Vespa 电动车的演变。
Digital sketch by Marco Canepa with the evolution of the Vespa design from early models to Vespa Elettrica.

比亚乔提供
Courtesy Piaggio

马尔科·卡内帕绘制的数字草图以及新款 Vespa 的侧视图。
Digital sketch by Marco Canepa with side view of the new Vespa.

比亚乔提供
Courtesy Piaggio

1945 年 12 月 2 日，工程师潘扎尼致工程师科拉迪诺·达斯卡尼奥的信件，内容涉及 Vespa MP3 原型草图和技术测试。
Letter with sketches from engineer Panzani to engineer Corradino D'Ascanio, concerning technical tests of the Vespa MP3 prototype, December 2, 1945.

维基共享资源 / 佩斯卡拉国家档案馆
Wikimedia Commons/Archivio di Stato di Pescara

## ⋀ 2020 年荣誉提名奖 Vespa 电动车

### 比亚乔集团设计中心——马尔科·兰布里和马尔科·卡内帕为比亚乔司设计，2018 年

### 电动摩托车

从自行车到滑板车再到电动车：在意大利乃至全球范围树立典范。

1946 年，前飞机制造商比亚乔对生产重新定位，Vespa 电动车由此诞生，并迅速成为全球轻型摩托车的典范，进行了大规模分销，取代了自行车和摩托车，充当汽车的平替。

在历经长达 80 年的发展后，Vespa 电动车最新改型，推出全电动版本，可以通过智能手机进行连接。

## ⋀ Honorable Mention 2020 Vespa elettrica

### Piaggio Group Design Center - Marco Lambri, Marco Canepa, Piaggio & C., 2018

### Electric motorcycle

From bicycles to scooters to electric mobility: setting a reference in Italy and worldwide.

Born in 1946 from a production repositioning of former aircraft manufacturer Piaggio, Vespa would soon become the worldwide reference for scooters, first with a large mass distribution, substituting bikes – and motorbikes – in providing an affordable alternative to cars.

After an uninterrupted 8-decade career, as an achievement in its most recent restylings, Vespa has also been released in a full-electric version that can connect to the user's smartphone.

2018 年 维布拉姆风吕敷布料鞋
维布拉姆公司设计，2016 年。
鞋类。
2018 Vibram Furoshiki The Wrapping Sole
Vibram, Vibram, 2016.
Footwear.

ADI 设计博物馆档案
Archivio ADI Design Museum

日本使用的风吕敷布料。
Furoshiki fabric in use in Japan.

细江烈档案
Isao Hosoe Archive

Big Footer 款的原型，打开和合上的样子。
Prototype of the Big Footer model, opened and closed.

细江烈档案
Isao Hosoe Archive

Shark 款的原型，打开和合上的样子。
Prototype of the Shark model, opened and closed.

细江烈档案
Isao Hosoe Archive

## ⋀ 2018 年 维布拉姆风吕敷布料鞋
### 维布拉姆公司设计，2016 年
### 鞋类

这是一款"布料"鞋，没有固定的形状，处于自然折叠状态，占用空间极小。

设计受日本传统方巾（风吕敷）的启发。风吕敷可以用于包裹提拿各种形状的物品。

风吕敷布料鞋的设计初衷是一款适合所有人的鞋，适用于任何脚型，就像方形的风吕敷，可以包容各种形状的物品。

## ⋀ 2018, Vibram Furoshiki The Wrapping Sole
### Vibram, Vibram, 2016
### Footwear

It is a 'textile' shoe, unstructured, which folds up on itself taking up minimal space.

Conceptual reference is the traditional square handkerchief (the furoshiki) that in Japanese culture is wrapped around objects to transport them by adapting to its shape.

Furoshiki is initially intended to respond to the idea of a one-size-fits-all shoe, adaptable to any foot, just as the square furoshiki fabric adapts to the shape of any object.

# 传播的力量

## The Power of Communication

获得金圆规奖的品牌与项目能够树立标志性形象，得益于强大的传播能力。通过广告宣传或现实生活场景，以及视觉文化中无孔不入的渗透，进入公众视野与脑海。可以说，金圆规奖首开先河，对优秀设计的价值予以认可，而这一价值又得到公众的肯定。人们购买优秀设计作品，认定它为完美诠释特定要求的象征，起到了很好的宣传作用。能够取得如此成绩，原因是多方面的。而品牌商对广告宣传的理解与重视，往往会将设计项目交由杰出的艺术家来创作完成。

Brands and projects that won the Compasso d'Oro have often established their iconic value through a strong communicational component, by establishing themselves within the common imaginary through legendary advertising campaigns or viral presence in real life situations and visual culture. We can say that the Compasso d'Oro award was the first to recognise the value of a 'well-done' design, a value that was then confirmed by the public, who bought it, contributing to its dissemination, or simply elected it as an icon as the interpreter of certain requirements. The reasons for this success are manifold and range from the immediate comprehensibility of designs such as the Eclisse lamp and the Solari flip boards, clear to everyone, to the apparent incomprehensibility of the Valentine typewriter, so deliberately different from its peers. These designs also owed their fortune to the far-sightedness of the manufacturers who understood the fundamental role of advertising campaigns, often entrusted to great artists as in the case of Campari.

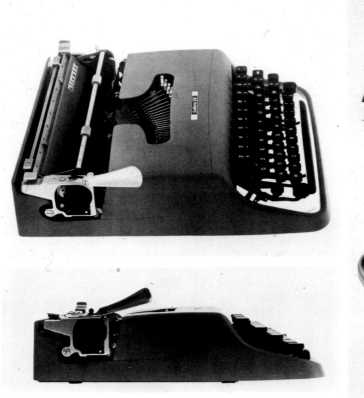

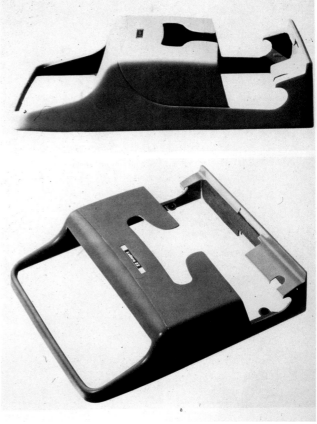

"字母 22" 打字机及其零部件的旧照。
Vintage photographs of Lettera 22 and its constituent parts

ADI 设计博物馆档案
Archivio ADI Design Museum

## 🔺 1954 年 "字母 22" 打字机

### 马尔切洛 · 尼佐利，Olivetti 公司，1950 年

### 打字机

有了 Olivetti 公司的便携打字机，无论是在办公室外，还是在家里，随时随地都可以工作。大众终于迎来了真正意义上的便携打字机："字母 22"。它小巧轻便，外壳设计为公文包样式，便于携带。

作为记者与作家的必备工具，又得益于铺天盖地的广告宣传，"字母 22"最终成为了大家争相抢购的"梦中情机"，也是意大利现代化的象征之一。

"字母 22"不再是专属于白领一族的办公用品，而是被包装成了送礼首选，是能够迅速传递思想，甚至是表达最私密想法的工具。因此，它成为了颇受顾客喜爱的情感传递工具，预示着如今人与通讯技术的紧密联系。

## 🔺 1954 Lettera 22

### Marcello Nizzoli, Olivetti, 1950

### Typewriter

Olivetti portable typewriters represent the possibility of working anywhere, outside the office or in our homes. Lightweight and transportable, the Lettera 22 is finally a true portable: its weight is low and its case in the form of a briefcase makes it easy to transport.

An indispensable tool for journalists and writers, it became, thanks to major advertising campaigns, an object of desire for everyone, one of the symbols of the modernisation of Italy.

The Lettera 22 no longer belongs only to the office world: it is presented as a perfect gift, as an instrument capable of quickly communicating thoughts, even the most personal ones. It thus becomes an emotional instrument of affection, a forerunner of today's link between man and communication technology.

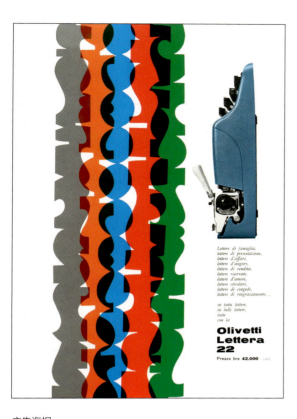

广告海报。
Advertising poster.

Olivetti 公司历史档案
Archivio Storico Olivetti

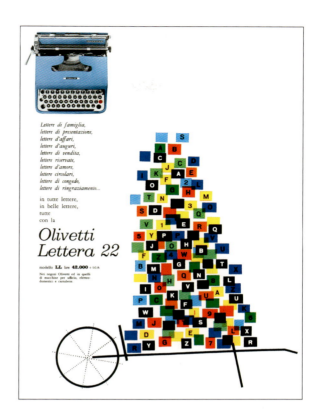

乔瓦尼·平托里设计的广告海报。
Advertising poster by Giovanni Pintori.

Olivetti 公司历史档案
Archivio Storico Olivetti

萨维尼亚克设计的平面广告。
Advertising graphics Raymond Savignac.

Olivetti 公司历史档案
Archivio Storico Olivetti

莱奥·廖尼设计的广告海报。
Advertising poster by Leo Lionni.

Olivetti 公司历史档案
Archivio Storico Olivetti

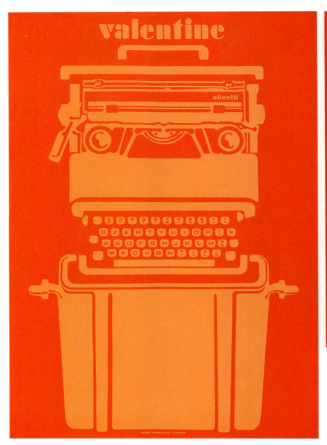

小埃托雷·索特萨斯和罗伯托·皮耶拉奇尼构思的广告海报。
Advertising poster conceived by Ettore Sottsass jr. and Roberto Pieracini.

Olivetti 公司历史档案
Archivio Storico Olivetti

## ⋀ 1970 年荣誉提名奖，"情人"打字机
### 小埃托雷 · 索特萨斯，Olivetti 公司，1969 年
### 便携打字机

最是那一抹耀眼的情人红……如果说打字机曾经像其他办公用品一样，放眼望去，不是深灰色，就是浅灰色，那么红色"情人"打字机的问世终于冲破了所有禁锢。

这款打字机选用塑料材质，有助于减轻重量，赋予使用者充分自由，将他们从办公室中解脱出来，不再受制于场所空间。

"情人"诞生于埃托雷 · 索特萨斯之手，他像其他知名设计师一样，设想了多个广告宣传方案；"情人"勇敢地诠释着所处时代的精神，真正为每一个人所打造：无论是飞行员还是年轻女性，无论是经理还是工人，都可以使用"情人"助力自己的工作与生活。

索特萨斯没有使用乏味的缩写代称，也没有使用枯燥的技术概念或功能性名称，而是通过采用引人遐思的女性名字来凸显这个手持装置的诱人魅力与其令人耳目一新的外观设计。尽管其销售成绩不尽人意，但它仍然是具有标志性意义的设计作品。

## ⋀ Honorable Mention 1970, Valentine
### Ettore Sottsass jr., Olivetti, 1969
### Portable typewriter

Valentine red. If typewriters, like other office objects, were previously dominated by the colour grey with a few concessions to pale tints, the red Valentine finally breaks all the mould.

'Plastic and Freedom' The material chosen helps to lighten this object, which finally leaves the office to conquer any place, any space.

Its creator, Ettore Sottsass, conceived several advertising campaigns, as did other important designers; the Valentine courageously interprets the spirit of its time, the Valentine is truly for everyone: for the airline pilot as for the young girl, for the manager and for the worker.

No longer a name that is an acronym, or one that refers to technical concepts or function, but an allusive woman's name that emphasises the seductiveness of this handset with its never-before-seen shapes. Despite its sales failure, a true icon.

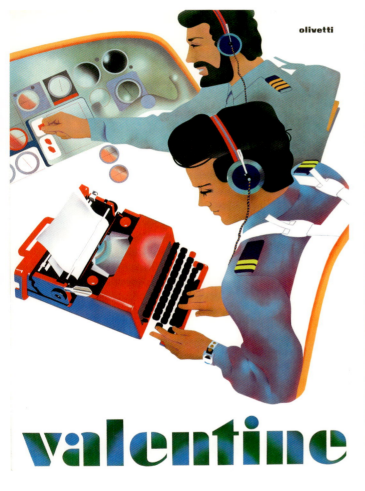

范德·埃尔斯特设计的广告海报。
Advertising poster by Adrianus van der Elst.

Olivetti 公司历史档案
Archivio Storico Olivetti

米尔顿·格拉泽尔设计的广告海报。
Advertising poster by Milton Glaser.

Olivetti 公司历史档案
Archivio Storico Olivetti

设备装置的旧照。
Vintage photograph of the mechanisms.

ADI 设计博物馆档案
Archivio ADI Design Museum

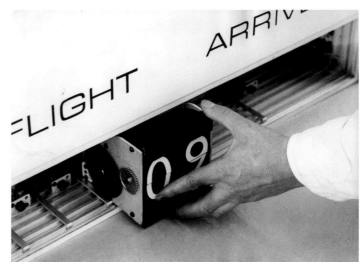

## ⋀ 1962 年 机场火车站显示设备

### 吉诺 · 瓦莱、Solari & C（Solari di Udine）公司 1956 年

### 翻牌式显示器

Solari 公司的翻牌式显示器完美融合技术巧思与有效传播，运行高效、清晰易懂、设计简洁精炼，能够为所有人提供所需信息。

它的设计师吉诺 · 瓦莱后续又为同一家公司设计了"数字5"（Cifra 5）翻牌式电子时钟，巧妙地将大型显示设备的设计迁移到了日常用品中，且不失功能性。

在世界各地的机场和火车站，翻页板不断地旋转组成准确的信息。随时间流逝，只要听到它们发出的"喀喀喀喀"的工作音，人们就能联想到旅客步履匆匆的画面。

## ⋀ 1962 Alpha-numeric remote indicators for airports and railway stations

### Gino Valle, Solari & C (Solari di Udine), 1956

### Split-flap display

Solari flip boards combines technique with communicative effectiveness: an efficient and clearly comprehensible system, at the same time elementary and refined, capable of providing information to everyone.

Designer Gino Valle, who will produce the Cifra 5 table clock for the same company, skilfully modulates the design from small to large scale, without ever losing effectiveness.

The mechanical noise of the split-flap paddles rotating to compose the exact information has over time become synonymous with arrivals and departures for travellers all over the world.

环球航空航站楼，埃罗·萨里宁，旧照。
TWA Terminal, Eero Saarinen, vintage photographs.

ADI 设计博物馆档案
Archivio ADI Design Museum

旧照。
Vintage photographs.

ADI 设计博物馆档案
Archivio ADI Design Museum

装置技术图纸。
Technical drawing of the mechanism.

ADI 设计博物馆档案
Archivio ADI Design Museum

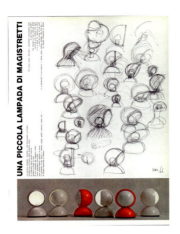

马吉斯特雷蒂的草图。
Sketches by Vico Magistretti.

*Domus* 杂志档案
© Editoriale *Domus* 联合股份有限公司
Archivio *Domus* © Editoriale *Domus* Spa

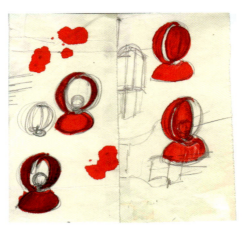

马吉斯特雷蒂的草图。
Sketches by Vico Magistretti.

维科 · 马吉斯特雷蒂基金会—马吉斯特雷蒂工作室档案
Archivio Studio Magistretti - Fondazione Vico
Magistretti

马吉斯特雷蒂的草图。
Sketches by Vico Magistretti.

维科 · 马吉斯特雷蒂基金会—马吉斯特雷蒂工作室档案
Archivio Studio Magistretti - Fondazione Vico
Magistretti

## ⋀ 1967 年"月食"台灯
### 维科 · 马吉斯特雷蒂，Artemide 品牌，1966/1967 年
### 台灯

这款台灯设计简洁，无需设计图纸，电话里就能解释清楚，设计师维科 · 马吉斯特雷蒂如是说。其设计灵感来自于维克多 · 雨果在小说《悲惨世界》中提及的暗灯，即用灯罩来遮蔽光源的灯。

它不仅设计简洁，就连名称也颇有意趣：月食（Eclisse）。当时，征服月球象征着人类的梦想，而这款台灯直接通过灯罩遮蔽面积的大小来体现月相，效果十分突出。

马吉斯特雷蒂以现代手法重新诠释了灯罩的形态，也反映出了颇具商业价值的理念，即人人都需要一盏床头灯。

## ⋀ 1967 Eclisse
### Vico Magistretti, Artemide, 1966/67
### Table lamp

A design so simple that it could be explained over the phone without the need for any design drawings, as its author Vico Magistretti calls it. The idea comes from the blind lantern, a lamp with a flap to obscure the light source, mentioned by Victor Hugo in his novel 'Les Miserables'.

The simplicity of the object is combined with the great narrative force of the name: Eclisse [Eclipse]. At a time when the conquest of the moon is a dream and a symbol, the direct reference between the different configurations that light can assume and the phases of the moon is highly effective.

Magistretti reinterprets the typology of the abat-jour in a modern key, also reflecting in a commercial sense: everyone needs a light on the bedside table next to the bed.

维科·马吉斯特雷蒂与各式"月食"台灯。
Vico Magistretti with various Eclisse lamps.

维科·马吉斯特雷蒂基金会—马吉斯特雷蒂工作室档案
Archivio Studio Magistretti - Fondazione Vico Magistretti

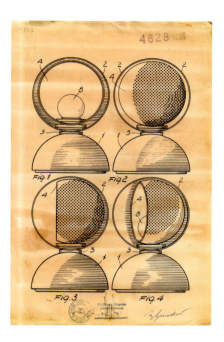

发明专利。
Patent of invention.

意大利国家中央档案馆
Archivio Centrale dello Stato

广告宣传。
Advertising campaign.

ADI 设计博物馆档案
Archivio ADI Design Museum

福尔图纳托·德佩罗，
《Paesaggio quasi tipografico》杂志，金巴利甜酒。
Fortunato Depero, Paesaggio quasi tipografico, Cordial Campari.

金巴利画廊
Galleria Campari

拉乌尔·沙雷温（笔名普里莫·西诺皮科）为金巴利苦味酒设计的作品，1920—1930 年。
Primo Sinòpico - Raoul Chareun- for Campari Bitter, 1920-30.

金巴利画廊
Galleria Campari

拉乌尔·沙雷温（笔名普里莫·西诺皮科）为金巴利苦味酒设计的作品，1920—1930 年。拼贴画。
Primo Sinòpico - Raoul Chareun- for Campari Bitter, 1920-30. Collage.

金巴利画廊
Galleria Campari

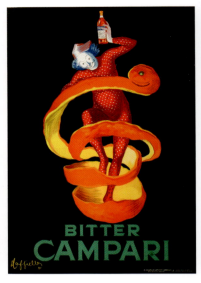

莱奥内托·卡皮耶洛设计的广告海报，《酒之精灵》，1921 年。
Advertising poster by Leonetto Cappiello, Lo Spiritello, 1921.

金巴利画廊
Galleria Campari

布鲁诺·穆纳里，《金巴利字样的多种平面设计图》，1964 年。
Bruno Munari, Declinazione Grafica del nome Campari, 1964.

金巴利画廊
Galleria Campari

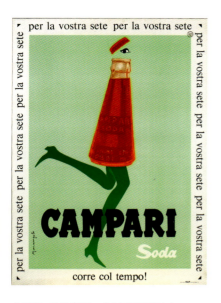

弗朗茨·马兰戈洛，《金巴利苏打水与
时俱进》，1960 年。
Franz Marangolo, Campari Soda corre
col tempo, 1960.

金巴利画廊
Galleria Campari

弗朗茨·马兰戈洛的拼贴画，《金巴利苏打水，我们永远的好伙伴》，1960 年。
Collage by Franz Marangolo, Campari Soda con noi sempre e ovunque, 1960.

金巴利画廊
Galleria Campari

## ⋀ 1991 年，金巴利酒 130 年图形史

**莱奥内托 · 卡皮耶洛、福尔图纳托 · 德佩罗、
马尔切洛 · 杜多维奇、布鲁诺 · 穆纳里、
乌戈 · 内斯波洛、马尔切洛 · 尼佐利等人，
达维德 · 坎帕里**

### 平面设计

设计经典标志：从未来派艺术家福尔图纳托 · 德佩罗设计的金巴利
苦味酒瓶开始，金巴利公司就比世界上其他许多公司都更早地认识到，
借助艺术家宣传其品牌形象是何等重要。

多年来，公司召集了无数设计大咖对公司形象加以诠释。当然，瓶身
设计始终是主打标志之一，"金巴利红"以及海报和宣传广告中的人
物形象也起到了辅助作用。

经过多年沉淀，金巴利已经成为一种象征，在世界各国眼中是"意式
风情"的代言人。从莱奥内托 · 卡皮耶洛的"酒之精灵"到布鲁诺 ·
穆纳里的"金巴利"字样平面设计作品，设计质量始终是缔造并支持
该品牌在国际上取得成功的最重要因素。

## ⋀ 1991, 130 Anni di storia della grafica Campari (130 years of Campari graphics)

**Leonetto Cappiello, Fortunato Depero, Marcello
Dudovich, Bruno Munari, Ugo Nespolo, Marcello
Nizzoli, et al., Davide Campari**

### Graphics

Designing an Icon: starting with the Campari bitter bottle,
designed by futurist artist Fortunato Depero, the company was
among the first in the world to understand the importance of
relying on artists to communicate its brand.

Over the years, numerous important figures have been called
upon to interpret the company's corporate identity. Certainly the
bottle remains one of the strongest symbols, together with the
colour 'Campari red', the protagonist of posters and advertising
campaigns.

Over the years, Campari has become a symbol, an icon
recognised throughout the world as an interpreter of the
'Italian way'. From Leonetto Cappiello's 'Spiritello' to the name
'Campari' interpreted by Bruno Munari, design quality has
always been the protagonist and architect of the brand's
international success.

# 工艺与工业

## Crafts and Industry

意大利设计文化凝聚了两种重要的影响因素：一是传统手工艺技术的传承；二是一以贯之的工业网络。设计界扎根于这一沃土，博采众长，重回意大利文化最根本的出发点——手工艺技术、材料与美学，从中汲取精华、挖掘灵感，推动工业与技术进步。而要解决工业领域提出的挑战，一方面可以依赖设计界在基本原型方面的深厚传承，另一方面还可以借助设计师与掌握复杂技术的小规模专业生产商之间的密切关系来协调解决问题。

要解决工业层面提出的挑战，一方面可以依赖设计领域在基本原型方面的深厚传承，另一方面可以借助设计师与能够掌握复杂技术的小规模专业生产商之间的密切关系。

Italian design culture brings together two very important influences: the heritage of traditional craft techniques and a consistent industrial network. Design has taken advantage of such a context, taking the best out of each production field and pushing industrial and technological advancements further by studying the essence of artisanal techniques, materials and esthetics which are a fundamental point of departure of Italian culture.

The challenges proposed by the industry can count on a strong design tradition of the fundamental archetypes and on the close relationship between designers and small specialized producers capable of developing complex techniques.

用弹簧秤吊起超轻椅的女人，20 世纪 50 年代。
Woman lifting a Superleggera chair with a spring
weight scale, 1950s.

威尼斯建筑大学，项目档案，乔治 · 卡萨利基金
Università Iuav di Venezia, Archivio Progetti, Fondo
Giorgio Casali

一名男童仅用小拇指就将超轻椅勾起，20 世纪 50 年代。
Boy lifting a Superleggera chair just with his little
finger, 1950s.

威尼斯建筑大学，项目档案，乔治 · 卡萨利基金
Università Iuav di Venezia, Archivio Progetti, Fondo
Giorgio Casali

Cassina 公司工厂，20 世纪 50 年代。
Cassina factory, 1950s.

Cassina 公司档案
Archivio Cassina

## ⚠ 1957 年荣誉提名奖，699 超轻椅
### 吉奥 · 蓬蒂，自 1957 年起由 Cassina 公司生产
### 椅

"超轻亦能承重，造型恰到好处，价格也十分低廉……这就是一把简简单单的椅子，朴实无华，无需修饰……"这款超轻椅的椅架使用白蜡木制作，椅子腿横截面为三角形，为尽可能减少用料，椅子腿自上而下逐渐收分。椅子仅重 1.7 千克，只用一根手指就可拎起，而且可以叠放。

通过超轻椅，蓬蒂对其原型——来自利古里亚海岸手工艺传统的 Chiavari 椅——重新加以诠释。他去除了所有矫揉造作的元素，摒弃了所有冗余的线条，实现了形式与结构同一的纯粹性。它轻盈（在第二版中更名为 Superleggera/ 超轻）、纤细，价格低廉，满足了面向大众的工业产品所必须达到的要求，且组装时无需螺丝，全靠接口，达到如此水平完全要归功于精湛的制作工艺。

"如果您去 Cassina 公司参观的话，他们会给您进行演示，让超轻椅做令人炫目的高空坠落，但椅子从不会摔碎，甚至还会反弹……走进他们的工厂，您可要小心，因为超轻椅一直在进行测试，在空中飞来飞去。"

## ⚠ Honorable Mention 1957, 699
### Gio Ponti, production Cassina since 1957
### Chair

"A chair that is light and strong at the same time, with the right shape, low price (...) A chair-chair, modestly, without adjectives (...) ". Made of ash wood, with triangular-section legs tapered to minimise material, this chair is easy to lift with one finger. It weighs 1.7 kg and is stackable.

Ponti reinterprets an archetype: the Chiavari chair from the Ligurian craft tradition. He frees it from every affectation, every redundant line, to achieve the purity of a form that identifies with the structure. Light (renamed Superleggera/Superlight in a second version), thin and cheap as an industrial product designed for everyone must be, but only realisable thanks to craftsmanship, it is assembled without screws, only with joints.

"If you go to Cassina they will show these chairs falling after dizzying flights, high and long, bouncing and never breaking (...) Entering their factory is dangerous because the chairs fly continuously in these incredible tests."

两个不同颜色花瓶的旧照。
Vintage photograph of two models in
different colours.

ADI 设计博物馆档案
Archivio ADI Design Museum

木制模具。
Wooden mould.

摄影：米莫·卡普尔索 /Saviati 工坊藏品 /ADI
设计博物馆档案
Photo by Mimmo Capurso / Collezione Salviati
/ Archivio ADI Design Museum

配有草图、照片和剖面图的说明文件。
Descriptive document with sketches,
photograph and section of the vase.

私人收藏
Private collection

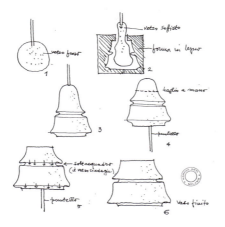

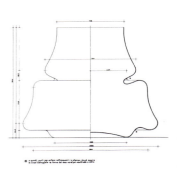

## ⋀ 1962 年 马尔科

### 塞尔焦 · 阿斯蒂，Salviati & C. 工坊，1962 年

### 花瓶

得益于形式研究与技术分析的共同助力，玻璃的机械形态达到新高，
本系列的各款花瓶即是实证。

塞尔焦 · 阿斯蒂的作品完美再现了吹制玻璃作品独有的流转延展，
以此与大众产生共鸣。

花瓶吹制时，首先需要通过木质模具塑形，然后再由人工进行热变形
的操作，从而改变了玻璃的原有形态。

## ⋀ 1962 Marco

### Sergio Asti, Salviati & C., 1962

### Flower vase

Formal research and technological analysis combine to achieve
an innovative formal solution for glass mechanically possible,
tested through the variations of the series.

Sergio Asti creates a widely distributed empathetic object,
where the flexible nature of glass, rooted in the tradition of
blown-glass artisans, becomes evident to a broad industrial
audience.

Initially formed in a wooden mould, the piece is then warped by
hand, which changes its original profile.

草图。
Sketch.

布卢默档案
Archivio Blumer

掀开椅面的木制椅架原型。
Opened-up prototype of the wooden structure.

布卢默档案
Archivio Blumer

## ⋀ 1998 年 Laleggera
### 里卡尔多 · 布卢默，Alias 品牌，1996 年
### 椅

Laleggera 这个名字显然是参考了吉奥 · 蓬蒂的作品。布卢默从中汲取灵感，探寻最为本真的形式，尽可能使用最少的材料，并由此减轻成品的重量。

在设计上，该作品最令人兴奋之处在于其饰面和内里所用材料的反差。传统木材装饰于外，典型的现代材料聚氨酯泡沫填充隐藏于内。坐在椅子上，人们只能感受到木材的温润。

甚至是在广告宣传方面，儿童用一根手指举起椅子的这一借用形象，也会让我们立刻想起吉奥 · 蓬蒂在推广超轻椅时所采用的相似画面。

## ⋀ 1998 Laleggera
### Riccardo Blumer, Alias, 1996
### Chair

The name is a clear reference to Gio Ponti's chair from which La leggera is inspired in the search for an essential form, in the use of the minimum possible material and in the consequent reduced weight.

The most stimulating element of the design is in the contrast between the traditional materials used on the outside and the interior, which instead employs a typically modern material, polyurethane foam, which remains concealed. Those using the chair only perceive the warmth of wood.

Even in the communication, the expedient of the child lifting the chair poised on a finger immediately brings us back to similar historical images conceived for Gio Ponti's Superleggera.

《精湛》，Foscarini 品牌，2018 年。
Maestrie, Foscarini, 2018.

Foscarini 品牌
Foscarini

使用光具座与即时成像胶片拍摄的照片，2018 年。
Photograph taken with an optical bench and instant film, 2018.

摄影：马西莫·加尔多内，马西莫·加尔多内提供。
Photo by Massimo Gardone, Courtesy Massimo Gardone

## ⋀ 2001 年  Tite，Mite

### 马克 · 萨德勒，Foscarini 品牌，2000 年

### 灯

"第四次工业革命呼唤意式设计复兴'人文主义精神'，不要仅把生产产品看作纯粹的技术挑战。"

手工组装技术原材料，很好地将手工艺技术的历史传承与最先进的工业发展成果融为一体。

组装人员充分发挥工匠精神，将玻璃布和一束不规则的碳纤维线缠绕在长长的钢制灯体上，兼顾了稳定性、美观性和散光性，赋予每一盏灯独一无二的气质。

## ⋀ 2001 Tite, Mite

### Marc Sadler, Foscarini Murano, 2000

### Lamps

"In the fourth industrial revolution, Italian design is called upon to renew its "humanism", going beyond the idea of production as a purely technological challenge."

Bare technological materials assembled by hand bring together the heritage of artisanal techniques and the achievements of state-of-the-art industrial developments.

Following the principles of craftsmanship, a fibreglass fabric and an irregular skein of carbon fibre thread are wound by hand around the steel shaft, solving at the same time the stability, the shape, and the diffusion of the light and making each lamp unique.

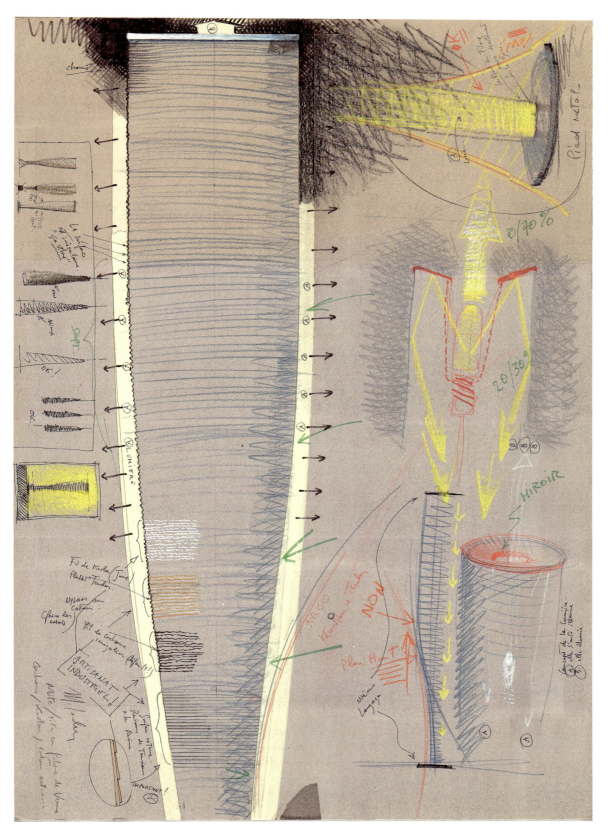

Mite 灯，草图，马克·萨德勒，2000 年。
Mite, sketch, Marc Sadler, 2000.

Sadler Associati 设计公司提供
Courtesy Sadler Associati

Klitter 吸音墙板，费利西娅·阿维兹，2018 年。
Klitter, Felicia Arvid, 2018.

费利西娅·阿维兹
Felicia Arvid

编织吸音布的原型，2018 年。
Prototype with woven fabric strips, 2018.

费利西娅·阿维兹
Felicia Arvid

夹子的原型和最终版，2018—2019 年。
Prototypes and final version of the clip,
2018-2019.

费利西娅·阿维兹和 Caimi Brevetti 公司
Felicia Arvid and Caimi Brevetti.

由透明亚克力夹子和木制加劲杆将垂直布条固定形成的墙板原型，2018年。
Prototype of wall panel conformed by vertical strips with clear acrylic clips and wooden stiffening rods, 2018.

费利西娅·阿维兹
Felicia Arvid

2022 年 Klipper
费利西娅·阿维兹，Caimi Brevetti 公司，
2019 年
隔音系统
2020 Klipper
Felicia Arvid, Caimi Brevetti, 2019
Soundproofing system

## ⋀ 2022 年 Klipper

### 费利西娅 · 阿维兹，Caimi Brevetti 公司，2019 年
### 隔音系统

将相同的模块反复折叠，就诞生了这款以简洁高效而闻名的隔音系统。

Klipper 是合作的产物：最初的手工原型灵感来自斯堪的纳维亚的山丘，而后续的制作则要归功于 Caimi Brevetti 公司长期以来在吸音布领域的丰富经验和先进技术。

在简洁高效的手工原型基础上，用金属夹将原型的两个组成部分夹紧固定，形成最终产品，但需要由用户在墙面上进行安装。

## ⋀ 2022 Klipper

### Felicia Arvid, Caimi Brevetti, 2019
### Soundproofing system

The repeated modular folds a product which is emblematic for its simplicity and effectiveness in the control of acoustics.

Klipper is the result of joint forces: the initial steps of artisanal prototypes inspired in the Scandinavian dunes and the highly technological advancements in the field of acoustic fabrics of an industrial company with a long-lasting experience in the field.

The simplicity and effectiveness of the handmade prototypes is translated to the final product which is produced with two patterns and hold together with metal clips and is meant to be mounted by the user.

# 时光印迹：
# 一场设计体验之旅

阿尔多·西比克  乔久园

# Articulating an
# Experience

Aldo Cibic and Joseph Dejardin

本次展览是意大利工业设计协会（ADI）和金圆规奖首次登陆中国，亮相上海地标外滩，进驻装饰艺术建筑明珠——外滩十八号。这幢老建筑气势恢宏、位置绝佳，虽经改造，但依然保有浓郁的历史文化气息；作为文化遗产保护的典范，完美契合举办如此一场经典设计展览的需要。旧日的优雅渗透在外滩十八号的当代设计创新中，融合成展示欣赏设计艺术与设计科学的"天选之所"。在这里，设计作品绽放光芒，设计领域的核心原则与创造力更是得以彰显。

展览沿时间线铺陈，展示自1954年奖项创立至今的历届金圆规奖获奖作品，每件展品均配有简短的获奖理由陈述。整条时间线细分为七个"分展区"，分别呈现不同主题，深入解析约45件展品。

各个分展区均以专属色彩凸显其迥异特质。展览运用墙壁、平台、桌、矮台等语素，构建横跨各分展区的统一设计语言；同时兼顾不同展陈空间的布展方式，贴合各分展区展品的主题特色。

外滩十八号是装饰艺术风格的历史风貌建筑，展览举办地位于大楼一、二层间夹层，可以俯瞰两层楼挑高的中庭。时间线围绕中庭空间流转，七大分展区沿时间轴线开枝散叶。

This exhibition, which marks ADI and the Compasso d'Oro prize's introduction in China, is being hosted at Bund 18, an impressive Art Deco architectural gem located on the Bund, Shanghai's most iconic waterfront promenade. Bund 18 is an ideal venue for hosting a design exhibition due to its historical significance, architectural grandeur, prime location, adaptive use, cultural fusion, and its capacity to showcase heritage conservation. The juxtaposition of historical elegance with contemporary design innovation creates a compelling environment for exhibiting and appreciating the art and science of design. It is a space that not only celebrates design but also embodies the core principles and creativity associated with this field.

The exhibition is designed around a timeline, showcasing the designs that have won the award from its foundation in 1954 until now. Each object on this timeline is described with a short statement explaining its relevance. Through the timeline are a series of seven "Insight" spaces that go in depth on approximately 45 objects, each exploring one particular thematic area.

Throughout the exhibition, we used colour as a tool to express the different personalities of each space. The language of the design uses walls and platforms, tables and plinths to build up a consistent language across the spaces, while subtly altering the display methodology in each different space to represent the themes of the exhibits.

The building itself is a historic Art Deco landmark. The exhibition takes place on the mezzanine of the building overlooking a central double-height space. This double-height space serves as a framework around which the exhibition unfolds. The timeline of objects wraps around it, with the Insight sections positioned within the flow of the timeline.

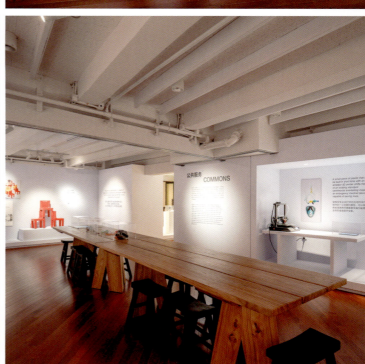

Λ 1998
Laleggera

Riccardo Blumer
Alias, 1996
Chair

1998 年
Laleggera
里卡尔多·布卢默
Alias 品牌，1996 年
椅

In 1953, La Rinascente proposed the exhibition L'estetica nel prodotto (Aesthetics in the Product), which served as a precursor to the Compasso d'Oro Award. For the first time, a department store addressed the "social problem" of the "greater dignity" of the product. The presentation brochure reads, "A knife, a dress, a glass, an iron, have their dignity not only in price or function but also in form (…)."

1953 年，文艺复兴百货公司提议举办题为"产品美学"的展览，即金圆规奖的前身。这是首次由一家百货公司来解决产品的"广义尊严"这个"社会问题"。展览宣传册中这样写道："无论是一把刀、一条裙子，还是一只玻璃杯、一个熨斗，产品的尊严与价值不仅体现在价格与功能上，也关乎外观形态……"

图书在版编目(CIP)数据

金圆规奖 : 引领意大利设计潮流七十年 / 意大利工业设计协会博物馆, 上海久事美术馆编. -- 上海 : 上海书画出版社, 2024.3

ISBN 978-7-5479-3316-9

Ⅰ. ①金… Ⅱ. ①意… ②上… Ⅲ. ①产品设计—作品集—意大利—现代 Ⅳ. ①TB472

中国国家版本馆CIP数据核字(2024)第043911号

**金圆规奖：引领意大利设计潮流七十年**

意大利工业设计协会博物馆　上海久事美术馆　编

| | |
|---|---|
| 责任编辑 | 王聪荟 |
| 审　读 | 王　剑 |
| 装帧设计 | 字面意义工作室 |
| 技术编辑 | 包赛明 |

| | |
|---|---|
| 出版发行 | 上 海 世 纪 出 版 集 团<br>上海书画出版社 |
| 地址 | 上海市闵行区号景路159弄A座4楼 |
| 邮政编码 | 201101 |
| 网址 | www.shshuhua.com |
| E-mail | shuhua@shshuhua.com |
| 印刷 | 浙江海虹彩色印务有限公司 |
| 经销 | 各地新华书店 |
| 开本 | 889×1194　1/16 |
| 印张 | 9.25 |
| 版次 | 2024年3月第1版　2024年3月第1次印刷 |
| 书号 | ISBN 978-7-5479-3316-9 |
| 定价 | 286.00元 |

若有印刷、装订质量问题，请与承印厂联系